GOVERNMENT PROCEDURES AND OPERATIONS

THE U.S. SCIENCE AND ENGINEERING WORKFORCE

EMPLOYMENT AND WAGE TRENDS

GOVERNMENT PROCEDURES AND OPERATIONS

Additional books in this series can be found on Nova's website under the Series tab.

Additional E-books in this series can be found on Nova's website under the E-book tab.

GOVERNMENT PROCEDURES AND OPERATIONS

THE U.S. SCIENCE AND ENGINEERING WORKFORCE

EMPLOYMENT AND WAGE TRENDS

RUSSELL B. HARTLEY
EDITOR

NOVINKA

New York

For permission to use material from this book please contact us:
Telephone 631-231-7269; Fax 631-231-8175
Web Site: http://www.novapublishers.com

LIBRARY OF CONGRESS CATALOGING-IN-PUBLICATION DATA

ISBN: 978-1-62948-932-2

Published by Nova Science Publishers, Inc. † New York

CONTENTS

PREFACE

Many congressional policymakers have maintained an ongoing interest in the adequacy of the number of U.S. scientists and engineers required to address the needs of U.S. employers, to spur economic growth and job creation through innovation, to maintain U.S. global technological leadership and industrial competitiveness, and to help address important national and societal needs. This book provides employment, wage, and unemployment information for the computer occupations, mathematical occupations, engineers, life scientists, physical scientists, and S&E management occupations. It also discusses immigration of foreign nationals with science, technology, engineering, and mathematics (STEM) degrees; and STEM education.

Chapter 1 – The adequacy of the U.S. science and engineering workforce has been an ongoing concern of Congress for more than 60 years. Scientists and engineers are widely believed to be essential to U.S. technological leadership, innovation, manufacturing, and services, and thus vital to U.S. economic strength, national defense, and other societal needs. Congress has enacted many programs to support the education and development of scientists and engineers. Congress has also undertaken broad efforts to improve science, technology, engineering, and math (STEM) skills to prepare a greater number of students to pursue science and engineering (S&E) degrees. Some policymakers have sought to increase the number of foreign scientists and engineers working in the United States through changes in visa and immigration policies.

Many policymakers, business leaders, academicians, S&E professional society analysts, economists, and others hold diverse views with respect to the adequacy of the S&E workforce and related policy issues. These issues include the question of the existence of a shortage of scientists and engineers in the United States, what the nature of such a shortage might be (e.g., too few people with S&E degrees, mismatched skills and needs), and whether the federal government should undertake policy interventions to address such a putative shortage or to allow market forces to work in this labor market. Among the key indicators used by labor economists to assess occupational labor shortages are employment growth, wage growth, and unemployment rates.

Chapter 2 – Although the United States remains the leading host country for international students in science, technology, engineering, or mathematics (STEM) fields, the global competition for talent has intensified. A record number of STEM graduates—both U.S. residents and foreign nationals—are entering the U.S. labor market, and there is a renewed focus on creating additional immigration pathways for foreign professional workers in STEM fields. Current law sets an annual worldwide level of 140,000 employment-based admissions, which includes the spouses and children in addition to the principal (i.e., qualifying) aliens. "STEM visa" is shorthand for an expedited immigration avenue that enables foreign nationals with graduate degrees in STEM fields to adjust to legal permanent resident (LPR) status without waiting in the queue of numerically limited LPR visas. The fundamental policy question is should the United States create additional pathways for STEM graduates to remain in the United States permanently?

The number of full-time graduate students in science, engineering, and health fields who were foreign students (largely on F-1 nonimmigrant visas) grew from 91,150 in 1990 to 148,923 in 2009, with most of the increase occurring after 1999. Despite the rise in foreign student enrollment, the percentage of STEM graduate students with temporary visas in 2009 (32.7%) was comparable to 1990 (31.1%). Graduate enrollments in engineering fields have exhibited the most growth of the STEM fields in recent years. About 40,000 graduate degrees were awarded to foreign STEM students in 2009, with 10,000 of those going to Ph.D. recipients.

Chapter 3 – The United States has traditionally produced the world's top research scientists and engineers, leading to breakthrough advances in science and technology. This technological innovation has been a primary driver of U.S. economic growth, with studies showing that half or more of economic growth in the United States over the past fifty years is attributable to improved productivity resulting from innovation. The late 20th century ushered in vast improvements in computer and information technologies, as well as biomedical technologies. These innovations are changing the way we live, work, and play in marvelous and unforeseen ways. Fully capturing the economic benefits of existing and undiscovered technologies will require a steady stream of Americans equipped with science, technology, engineering, and math (STEM) knowledge, skills, and abilities.

The need for workers with STEM skills is heightened in today's global economy. Technological innovation improves the competitive position of U.S. industries, drives export growth, and supports high-quality jobs. Additionally, demand for STEM-capable workers has increased even in traditionally non-STEM fields due to the diffusion of technology across industries and occupations. Technological progress can improve workers' quality of life by improving working conditions and increasing wages. However, for workers to benefit from the economic gains generated by innovation, they must have the knowledge and skills needed to work effectively in jobs requiring STEM competencies. Improving access to quality STEM education will strengthen the caliber of the U.S. workforce, drive economic growth, and keep the U.S. competitive.

In: The U.S. Science and Engineering Workforce ISBN: 978-1-62948-932-2
Editors: Russell B. Hartley

Chapter 1

THE U.S. SCIENCE AND ENGINEERING WORKFORCE: RECENT, CURRENT, AND PROJECTED EMPLOYMENT, WAGES, AND UNEMPLOYMENT*

John F. Sargent

SUMMARY

The adequacy of the U.S. science and engineering workforce has been an ongoing concern of Congress for more than 60 years. Scientists and engineers are widely believed to be essential to U.S. technological leadership, innovation, manufacturing, and services, and thus vital to U.S. economic strength, national defense, and other societal needs. Congress has enacted many programs to support the education and development of scientists and engineers. Congress has also undertaken broad efforts to improve science, technology, engineering, and math (STEM) skills to prepare a greater number of students to pursue science and engineering (S&E) degrees. Some policymakers have sought to increase the number of foreign scientists and engineers working in the United States through changes in visa and immigration policies.

---

* This is an edited, reformatted and augmented version of a Congressional Research Service Publication, CRS Report for Congress R43061, dated May 6, 2013.

Many policymakers, business leaders, academicians, S&E professional society analysts, economists, and others hold diverse views with respect to the adequacy of the S&E workforce and related policy issues. These issues include the question of the existence of a shortage of scientists and engineers in the United States, what the nature of such a shortage might be (e.g., too few people with S&E degrees, mismatched skills and needs), and whether the federal government should undertake policy interventions to address such a putative shortage or to allow market forces to work in this labor market. Among the key indicators used by labor economists to assess occupational labor shortages are employment growth, wage growth, and unemployment rates.

In 2011, there were 5.9 million scientists and engineers employed in the United States, accounting for 4.6% of total U.S. employment. Science and engineering employment was concentrated in two S&E occupational groups, computer occupations (56%) and engineers (25%), with the rest accounted for by S&E managers (9%), physical scientists (4%), life scientists (4%), and those in mathematical occupations (2%). From 2008 to 2011 S&E employment increased by 99,550, rising to 5.9 million, a compound annual growth rate (CAGR) of 0.6%, while overall U.S. employment contracted at 1.7% CAGR. Viewed only in aggregate, the overall increase in S&E employment masks the varied degrees of growth and decline in the detailed S&E occupations.

In 2011, the mean wage for all scientists and engineers was $85,700, while the mean wage for all other occupations was $43,300. Between 2008 and 2011, the mean wages of each S&E occupational group grew more slowly (1.5%-2.2% CAGR) than the mean wage for all occupations (2.3% CAGR).

Compared to the overall workforce, the S&E occupational groups had significantly lower unemployment rates for the 2008-2011 period. In general, though, the professional occupations (of which the S&E occupations are a part) historically have had lower unemployment rates than the workforce as a whole. In 2011, the overall S&E unemployment rate of 3.9% was higher than for other selected professional occupations, including lawyers (2.1%), physicians and surgeons (0.6%), dentists (0.7%), and registered nurses (2.0%).

The Bureau of Labor Statistics projects the number of science and engineering jobs (as defined in this report) will grow by 1.1 million between 2010 and 2020, a growth rate (1.7% CAGR) that is somewhat faster than that of the overall workforce (1.3%). In addition, BLS projects that a further 1.3 million scientists and engineers will be needed to replace those projected to exit S&E occupations. Growth in the S&E occupational groups is projected to range from 1.0%-2.0% CAGR. The number of scientists and engineers needed to meet growth and net replacement needs between 2010 and 2020 is 2.4 million, including 1.4 million in the computer occupations and 525,900 engineers.

OVERVIEW

Many congressional policymakers have maintained an ongoing interest in the adequacy of the number of U.S. scientists and engineers required to address the needs of U.S. employers, to spur economic growth and job creation through innovation, to maintain U.S. global technological leadership and industrial competitiveness, and to help address important national and societal needs.

To help ensure an adequate S&E workforce, Congress has enacted and appropriated funds for a variety of federal programs. These programs intend to foster improved science, technology, engineering, and mathematics (STEM) skills among students; to incentivize students to pursue degrees in science and engineering (S&E) through tools such as fellowships, assistantships, and traineeships; and to provide graduate and post-graduate research experiences at U.S. colleges and universities through the financing of university-based research. The 113th Congress is considering legislation to create, reform, and provide funding for STEM education efforts, and may seek to reauthorize the America COMPETES Act of 2007 (P.L. 110-69) and the America COMPETES Reauthorization Act of 2010 (P.L. 111-358).[1] In addition, Congress is considering changes to immigration policies, among them the number and processes associated with F-1 visas, H-1B visas, L1 visas, and legal permanent residency (commonly referred to as "Green Cards"), to address U.S. S&E workforce needs.[2]

As Congress develops policies and programs and makes appropriations to help address the nation's needs for scientists and engineers, it may wish to consider past, current, and projected S&E workforce trends. In this regard, this report provides employment, wage, and unemployment information[3] for the computer occupations, mathematical occupations, engineers, life scientists, physical scientists, and S&E management occupations, in three sections:

- "Current Employment, Wages, and Unemployment" provides a statistical snapshot of the S&E workforce in 2011 (the latest year for which data are available) with respect to occupational employment, wage, and unemployment data.

- "Recent Trends in Employment, Wages, and Unemployment" provides a perspective on how S&E employment, wages, and unemployment have changed during the 2008-2011 period.
- "Employment Projections, 2010-2020" provides an analysis of the Bureau of Labor Statistics' occupational projections examining how the number employed in S&E occupations are expected to change during the 2010-2020 period, as well as how many openings will be created by workers exiting each occupation (replacement needs).

A final section, "Concluding Observations," provides various stakeholder perspectives that Congress may wish to consider as it seeks to ensure that the United States has an adequate S&E workforce to meet the demands of the 21st century.

METHODOLOGY

Occupational Taxonomy

Most experts agree that there is no authoritative definition of which occupations comprise the science and engineering (S&E) workforce. Rather, the selection of occupations included in any particular analysis of the S&E workforce may vary. Some analysts, policymakers, and organizations may refer to the group in different ways (e.g., the scientific and technical workforce, the STEM workforce) and include varying sets of occupations. In 2001, the Bureau of Labor Statistics (BLS), in defining the STEM occupations for a particular analysis, stated, "This is only one possible definition of STEM occupations; other definitions exist that may be better suited for other uses."[4]

The size of the S&E workforce varies substantially depending on which occupations are included in the definition. In its 2012 Science and Engineering Indicators report, the National Science Board (NSB) stated, "In the most recent estimates, the U.S. S&E workforce (defined by occupation) totaled between 4.8 million and 6.4 million people."[5] Previously, the NSB asserted that the S&E workforce could be as large as 21 million people if the definition included those with either an S&E degree or a degree in an S&E-related field such as health or technology.[6]

The policy debate about the adequacy of the U.S. S&E workforce has focused largely on the computer occupations, mathematical occupations, engineers, and physical scientists. For purposes of this report, these occupations, along with life scientists (a part of the natural sciences, with physics and chemistry) and S&E management occupations, are collectively referred to as the S&E workforce. Notably, this group does not include social scientists (e.g., economists, survey researchers, psychologists, sociologists, urban and regional planners, anthropologists, archeologists, geographers, historians, political scientists) or S&E-related technicians. As defined this way, the size of the S&E workforce in 2011 was approximately 5.9 million.

This report uses a modified version of the Standard Occupation Classification (SOC) system[7] to categorize scientists and engineers. The report taxonomy includes six S&E occupational groups, each composed of closely related detailed occupations:

- **computer occupations—**including computer and information research scientists; computer systems analysts; computer programmers; software developers, applications; software developers, systems software; database administrators; network and computer systems administrators; computer support specialists; information security analysts, web developers, and computer network architects; and computer occupations, all other;
- **mathematical occupations—**actuaries; mathematicians; operations research analysts; statisticians; and mathematical science occupations, all other;
- **engineers—**including aerospace, agricultural, biomedical, chemical, civil, computer hardware, electrical, electronics (except computer), environmental, health and safety (except mining safety engineers and inspectors), industrial, materials, mechanical, mining and geological (including mining safety engineers), nuclear, and petroleum engineers; engineers, all other; and marine engineers and naval architects;
- **life scientists—**animal scientists, food scientists and technologists, soil and plant scientists, biochemists and biophysicists, microbiologists, zoologists and wildlife biologists, biological scientists, all other, conservation scientists, foresters, epidemiologists, medical scientists (except epidemiologists), and life scientists, all other.

- **physical scientists—**astronomers, physicists, atmospheric and space scientists, chemists, materials scientists, environmental scientists and specialists (including health), geoscientists (except hydrologists and geographers), hydrologists, and physical scientists, all other; and
- **S&E managers—**computer and information systems managers, architectural and engineering managers,[8] and natural sciences managers.

A description of the detailed occupations is provided inn the **Appendix**.

Data Sources

This report relies on federal government employment, wage, and unemployment data from the following sources:

- The **Occupational Employment Statistics (OES)**,[9] a survey of non-farm establishments conducted by the Department of Labor's Bureau of Labor Statistics and state workforce agencies, is the source of employment and wage data for the 2008-2011 period. The survey provides employment and wage estimates annually for over 800 occupations. According to BLS, "employees" are all part-time and full-time workers who are paid a wage or salary. The survey does not cover the self-employed, owners and partners in unincorporated firms, household workers, or unpaid family workers. For this report, the wage statistic used is the occupational "mean wage," an average wage calculated by summing the wages of all the employees in a given occupation and then dividing the total wages by the number of employees.
- The **Current Population Survey (CPS)**,[10] a monthly survey of households conducted for BLS by the Department of Commerce's Bureau of the Census, is the source of the unemployment data in this report. CPS data are also used to supplement OES data in BLS employment projections (discussed below).
- BLS's **Employment Projections**,[11] a biennial product of BLS, provides occupational employment and industry employment projection data for ten-year periods. The latest projections, covering the 2010-2020 period, were published in January 2012. According to BLS, for most industries, the OES survey provides data for the

occupational staffing patterns—the distribution of wage and salary employment by occupation in each industry—and Current Employment Statistics (CES) [12] data provide information on total wage and salary employment in each non-farm industry. While OES data include only wage and salary, non-farm employment, the employment data in the projections also include agricultural industry employment and the self-employed (derived from CPS data) to arrive at 2010 employment levels for each occupation.[13]

Timeframe

The "Current Trends in Employment, Wages, and Unemployment" section provides information on changes in employment, wages, and unemployment for the period 2008 to 2011. The "Employment Projections, 2010-2020" section relies entirely on the most recent Bureau of Labor Statistics biennial employment projections for the 2010-2020 timeframe.

Methodological Limitations

A variety of factors may affect the comparability of OES data over time:

> Although the OES survey methodology is designed to create detailed cross-sectional employment and wage estimates for the U.S., States, metropolitan and nonmetropolitan areas, across industry and by industry, it is less useful for comparisons of two or more points in time. Challenges in using OES data as a time series include changes in the occupational, industrial, and geographical classification systems, changes in the way data are collected, changes in the survey reference period, and changes in mean wage estimation methodology, as well as permanent features of the methodology.[14]

In its examination of current trends, CRS chose the 2008-2011 time period to enhance comparability of data across the period by reducing inconsistencies resulting from changes in the OES occupational classification system. The OES survey used the same occupational categories throughout the 2008-2011 period.

The Bureau of Labor Statistics makes a number of estimates in developing its employment projections. These estimates include "the future size and composition of the population, as well as on the trends in labor force participation rates of different age, gender, race, and ethnic groups, a total of 136 separate categories," "the rate of growth and demand composition of real GDP, the labor productivity growth rate, and the inflation rate ," expectations regarding the federal budget surplus or deficit, historical staffing patterns, shifts in product mix, changes in technology and business practices, and retirement rates.[15] If these estimates do not accurately reflect future performance, occupational employment projections may be over- or underestimated. Other factors may affect occupational projections as well, including changes to immigration laws and patterns, trade laws and practices, regulatory regimes, and social and educational patterns; wars and disasters; revolutionary advances in technology; and shifts in consumer tastes. The BLS evaluates the accuracy of its projections regularly and publishes these evaluations in its *Monthly Labor Review*.[16]

SELECTED S&E OCCUPATIONAL DATA

Current Employment, Wages, and Unemployment

This section provides a snapshot of the S&E workforce in 2011, using employment, wages, and unemployment data.

Employment

In 2011, the latest year for which Occupational Employment Survey data are available, 5.9 million people were employed in the United States as scientists and engineers, accounting for 4.6% of total U.S. employment. Science and engineering employment was concentrated in two occupational groups—computer occupations and engineers—which together accounted for 81% of S&E jobs, with 56% and 25%, respectively. The remainder of S&E employment was accounted for by science and engineering managers (9%), physical scientists (4%), life scientists (4%), and mathematical occupations (2%). Employment totals and share of S&E occupational employment are presented in **Figure 1**.

Table 4 provides 2011 employment data for each of the individual S&E occupations, organized by S&E group.

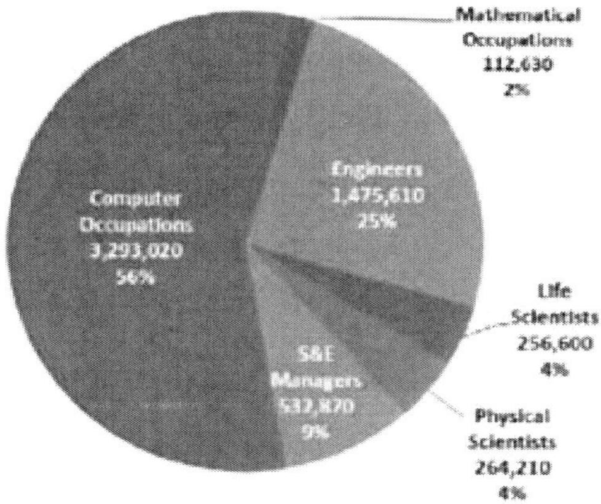

Source: CRS analysis of Occupational Employment Survey data, May 2011, Bureau of Labor Statistics, U.S. Department of Labor, http://www.bls.gov/oes/tables.htm.
Notes: Chart does not include social scientists or S&E-related technicians.

Figure 1. Share of S&E Occupational Employment, 2011.

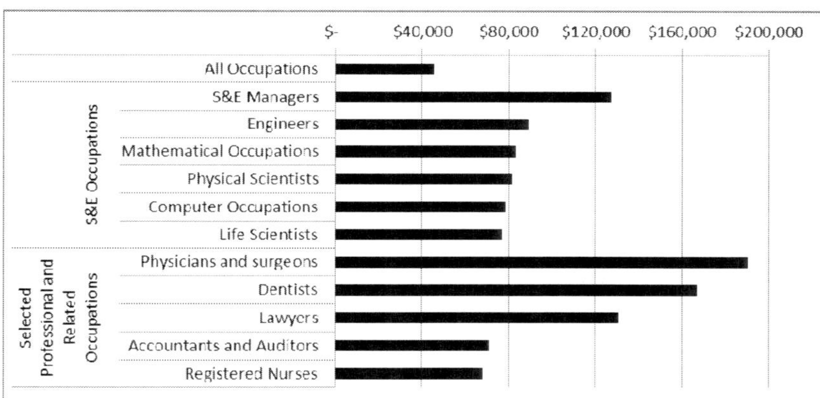

Source: CRS analysis of Occupational Employment Survey data, May 2011, Bureau of Labor Statistics, U.S. Department of Labor, http://www.bls.gov/oes/tables.htm.

Figure 2. Mean Wages of S&E Occupational Groups and Other Selected Professional Occupations, 2011.

Wages

Scientists and engineers, in general, have mean wages that exceed the mean wage for all occupations in the United States. In 2011, the mean wage for all occupations—professional and non-professional—was $45,230. S&E managers had the highest mean wage of all S&E occupational groups at $127,200, followed by engineers, $89,000; mathematical occupations, $81,400; computer occupations, $78,600; and life scientists, $76,800. Scientists and engineers earn less than some other professionals, such as physicians and surgeons ($190,100), dentists ($166,900), and lawyers ($130,900). (See **Figure 2.**)

Table 1 provides the 2011 mean wages for each of the individual S&E occupations, organized by S&E group.

Table 1. Mean Wages of S&E Occupations, 2011

Occupational Group	Mean Wage
Science and Engineering Managers	
Computer and Information Systems Managers	$125,660
Architectural and Engineering Managers	129,350
Natural Sciences Managers	128,230
Computer Occupations	
Computer and Information Research Scientists	$103,160
Computer Systems Analysts	82,320
Computer Programmers	76,010
Software Developers, Applications	92,080
Software Developers, Systems Software	100,420
Database Administrators	77,350
Network and Computer Systems Administrators	74,270
Computer Support Specialists	51,820
Information Security Analysts, Web Developers, and Computer Network Architects	81,670
Computer Occupations, All Other	80,500
Mathematical Occupations	
Actuaries	$103,000
Mathematicians	101,320
Operations Research Analysts	78,840
Statisticians	77,280
Mathematical Science Occupations, All Other	63,170
Engineers	
Aerospace Engineers	$103,870
Agricultural Engineers	78,400
Biomedical Engineers	88,360

Occupational Group	Mean Wage
Chemical Engineers	99,440
Civil Engineers	82,710
Computer Hardware Engineers	101,360
Electrical Engineers	89,200
Electronics Engineers, except Computer	94,670
Environmental Engineers	83,340
Health and Safety Engineers, except Mining Safety Engineers and Inspectors	78,540
Industrial Engineers	79,840
Marine Engineers and Naval Architects	91,730
Materials Engineers	86,790
Mechanical Engineers	83,550
Mining and Geological Engineers, Including Mining Safety Engineers	90,070
Nuclear Engineers	105,160
Petroleum Engineers	138,980
Engineers, All Other	92,260
Life Scientists	
Animal Scientists	$74,170
Food Scientists and Technologists	64,170
Soil and Plant Scientists	63,890
Biochemists and Biophysicists	87,640
Microbiologists	71,720
Zoologists and Wildlife Biologists	61,880
Biological Scientists, All Other	73,050
Conservation Scientists	62,290
Foresters	56,130
Epidemiologists	69,660
Medical Scientists, except Epidemiologists	87,640
Life Scientists, All Other	74,220
Physical Scientists	
Astronomers	$101,630
Physicists	112,090
Atmospheric and Space Scientists	90,860
Chemists	74,780
Materials Scientists	86,600
Environmental Scientists and Specialists, Including Health	68,810
Geoscientists, except Hydrologists and Geographers	97,700
Hydrologists	79,070
Physical Scientists, All Other	96,290

Source: CRS analysis of Occupational Employment Survey data, May 2011, Bureau of Labor Statistics, U.S. Department of Labor, http://www.bls.gov/oes/tables.htm.

Unemployment

The 2011 unemployment rates for S&E occupational groups are shown in **Figure 3**, together with the unemployment rates of other selected professional and related occupations.[17] The unemployment rates for the S&E occupations overall (3.9%) and for each individual S&E occupational groups (which range from 2.3% to 5.3%) were lower than the overall unemployment rate for those 16 and over (8.9%), and generally lower than the unemployment rate for the professional and related occupations group (4.3%).[18] However, the unemployment rates for the S&E occupational groups were higher than the rates for some other professional occupations, including lawyers (2.1%), secondary school teachers (3.0%), physicians and surgeons (0.6%), dentists (0.7%), and registered nurses (2.0%).

The unemployment rate for each detailed S&E occupation is provided in **Table 2.**[19] The unemployment rates for S&E occupations ranges from 1.2% for environmental scientists and geoscientists to 7.3% for computer support specialists.[20]

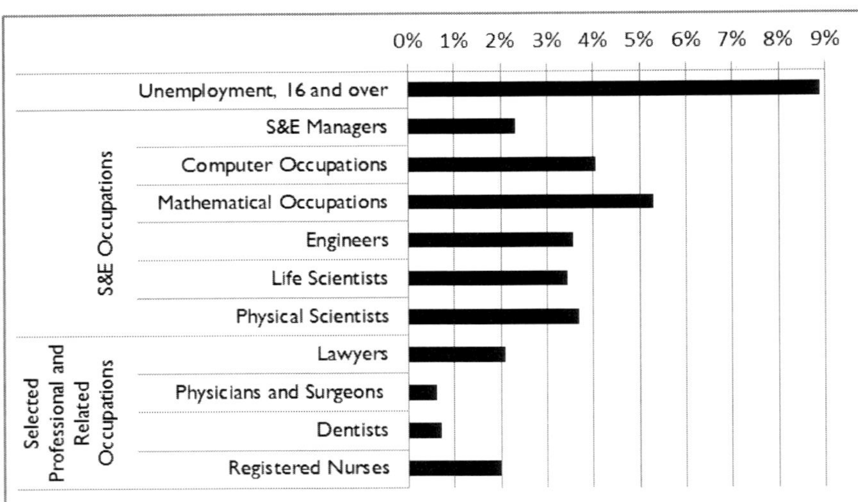

Source: CRS analysis of unpublished 2011 data from the Current Population Survey, Bureau of Labor Statistics.

Figure 3. Unemployment Rates for S&E Occupational Groups and Selected Professional and Related Occupations, 2011.

Table 2. Unemployment Rate for S&E Occupational Groups and Detailed S&E Occupations, 2011

Occupation	Unemployment Rate (Percentage)
Total, all occupations, 16 years and over	**8.9**
Science and Engineering Managers	**2.3**
Computer and information systems managers	2.9
Architectural and engineering managers	1.6
Natural sciences managers	n/a
Computer Occupations	**4.1**
Computer and information research scientists	n/a
Computer systems analysts	2.5
Information security analysts	n/a
Computer programmers	3.7
Software developers, applications and systems software	4.0
Web developers	4.7
Computer support specialists	7.3
Database administrators	1.3
Network and computer systems administrators	3.9
Computer network architects	0.4
Computer occupations, all other	4.7
Mathematical Occupations	**5.3**
Actuaries	n/a
Mathematicians	n/a
Operations research analysts	6.5
Statisticians	n/a
Miscellaneous mathematical science occupations	n/a
Engineers	**3.6**
Aerospace engineers	1.9
Agricultural engineers	n/a
Biomedical engineers	n/a
Chemical engineers	3.6
Civil engineers	4.8
Computer hardware engineers	2.3
Electrical and electronics engineers	3.4
Environmental engineers	n/a
Industrial engineers, including health and safety	5.5
Marine engineers and naval architects	n/a
Materials engineers	n/a

Table 2. (Continued)

Occupation	Unemployment Rate (Percentage)
Mechanical engineers	2.4
Mining and geological engineers, including mining safety	n/a
Nuclear engineers	n/a
Petroleum engineers	n/a
Engineers, all other	2.9
Life Scientists	**3.4**
Agricultural and food scientists	n/a
Biological scientists	2.9
Conservation scientists and foresters	n/a
Medical scientists	3.4
Life scientists, all other	n/a
Physical Scientists	**3.7**
Astronomers and physicists	n/a
Atmospheric and space scientists	n/a
Chemists and materials scientists	6.1
Environmental scientists and geoscientists	1.2
Physical scientists, all other	4.0

Source: Current Population Survey, 2011, Bureau of Labor Statistics, U.S. Department of Labor.

Notes: Unemployment for occupations refers to the experienced unemployed (those with prior work experience), classified according to their last job. For occupations in which the total number of employed and unemployed totals less than 50,000 the unemployment rate is not shown; n/a indicate data are not available.

Recent Trends in Employment, Wages, and Unemployment

This section provides information on changes in employment, wages, and unemployment for the period 2008 to 2011.

Employment Trends

Aggregate S&E Employment

During the 2008-2011 period, aggregate S&E employment increased by 99,550 jobs, rising from 5.8 million to 5.9 million, a compound annual growth rate of 0.6%.[21] The growth in the S&E occupations ran counter to overall U.S.

employment which contracted at 1.7% CAGR during this same period. Growth rates for the S&E occupational groups and detailed occupations are provided in the following sections.

Science and Engineering Occupational Groups

Table 3 provides employment data—2008 employment, 2011 employment, and changes in number employed and the compound annual growth rates during the 2008 to 2011 period—for each S&E occupational group. The fastest growth rate among S&E occupational groups during this period was in science and engineering managers which grew at 2.0% CAGR, while the largest increase in the number employed was in computer occupations which added 94,970 jobs. Among the S&E occupational groups, only engineers experienced a decline in employment, losing a total of 40,620 jobs during this period (-0.9% CAGR).

Detailed S&E Occupations

Table 4 provides 2008-2011 employment data for each of the S&E occupations, organized by S&E group. The data indicate that there was substantial variation in the number of jobs gained and lost among the S&E occupations, as well as in their growth rates. With respect to the number employed, the occupation with the largest gain was computer support specialists which added 86,970 jobs, while the occupation experiencing the largest decrease was computer programmers which lost 74,130 jobs. The S&E occupation with the fastest growth rate was astronomers with a 17.6% CAGR (though the number of new jobs (800) was small compared to other S&E occupations), while the occupation with the fastest decline was mathematical science occupations, all other, which experienced a -42.4% CAGR.

Among the computer occupations, those with the fastest growth rates were information security analysts, web developers, and computer network architects (5.8% CAGR); computer support specialists (5.1% CAGR); and computer software developers, applications (3.0% CAGR). These occupations also accounted for the vast majority of the job growth in the computer occupations group. These gains offset losses in other computer occupations, including computer programmers (-74,130, -6.7% CAGR); computer occupations, all other (-14,150, -2.5% CAGR); and database administrators (-7,270, -2.1% CAGR).

Several engineering occupations increased employment during this period, including aerospace engineers (11,600, 5.4% CAGR), petroleum engineers (10,000, 13.9% CAGR), nuclear engineers (1,790, 3.5%), mechanical engineers (4,650, 0.7% CAGR), and biomedical engineers (1,370, 2.9 % CAGR). Employment gains in these engineering occupations were offset by declines in twelve engineering occupations, including engineers, all other (-43,650, -9.5% CAGR); civil engineers (-7,230, -0.9% CAGR); electronics engineers (-3,620, -0.9% CAGR); industrial engineers (-3,090, -0.5%), and chemical engineers (-3,110, -3.5% CAGR).

Growth in the mathematical occupations was led by operations research analysts (4,170, 2.2% CAGR), statisticians (3,090, 4.8%), and actuaries (1,370, 2.4% CAGR). Mathematicians grew somewhat (210, 2.5% CAGR), while mathematical science occupations, all other declined (-5,340, -42.4% CAGR).

Among life scientists, the occupation biological scientists, all other, had the largest employment growth (3,260, 3.7% CAGR), while conservation scientists had the fastest growth rate (3,080, 6.1% CAGR). Employment declined in four life science occupations: medical scientists (-4,530, -1.5% CAGR); life scientists, all other (-2,010, -5.9% CAGR); foresters (-1,160, -4.0% CAGR); and animal scientists (-570, -7.4% CAGR).

Table 3. Employment Change in S&E Occupational Groups, 2008-2011

Occupation	Employment, 2008	Employment, 2011	Employment Change, Number	Employment Change, Compound Annual Growth Rate
All occupations	135,185,230	128,278,550	-6,906,680	-1.7%
All S&E occupations	5,835,390	5,934,940	99,550	0.6%
S&E managers	502,180	532,870	30,690	2.0%
Computer occupations	3,198,050	3,293,020	94,970	1.0%
Mathematical occupations	109,130	112,630	3,500	1.1%
Engineers	1,516,230	1,475,610	-40,620	-0.9%
Life scientists	250,250	256,600	6,350	0.8%
Physical scientists	259,550	264,210	4,660	0.6%

Source: CRS analysis of Occupational Employment Survey data, 2008-2011, Bureau of Labor Statistics, U.S. Department of Labor, http://www.bls.gov/oes/tables.htm.

Table 4. Employment in Detailed S&E Occupations, 2008-2011

Occupational Group	Employment, 2008	Employment, 2011	Employment Change, Number	Employment Change, Compound Annual Growth Rate
Science and Engineering Managers				
Computer and Information Systems Managers	276,820	300,830	24,010	2.80%
Architectural and Engineering Managers	182,300	184,530	2,230	0.40%
Natural Sciences Managers	43,060	47,510	4,450	3.30%
Computer Occupations				
Computer and Information Research Scientists	26,610	25,160	-1,450	-1.90%
Computer Systems Analysts	489,890	487,740	-2,150	-0.10%
Computer Programmers	394,230	320,100	-74,130	-6.70%
Software Developers, Applications	494,160	539,880	45,720	3.00%
Software Developers, Systems Software	381,830	387,050	5,220	0.50%
Database Administrators	115,770	108,500	-7,270	-2.10%
Network and Computer Systems Administrators	327,850	341,800	13,950	1.40%
Computer Support Specialists	545,520	632,490	86,970	5.10%
Information Security Analysts, Web Developers, and Computer Network Architects	230,410	272,670	42,260	5.80%
Computer Occupations, All Other	191,780	177,630	-14,150	-2.50%
Mathematical Occupations				
Actuaries	18,220	19,590	1,370	2.40%
Mathematicians	2,770	2,980	210	2.50%
Operations Research Analysts	60,860	65,030	4,170	2.20%
Statisticians	20,680	23,770	3,090	4.80%
Mathematical Science Occupations, All Other	6,600	1,260	-5,340	-42.40%
Engineers				
Aerospace Engineers	67,800	79,400	11,600	5.40%

Table 4. (Continued)

Occupational Group	Employment, 2008	Employment, 2011	Employment Change, Number	Employment Change, Compound Annual Growth Rate
Agricultural Engineers	2,640	2,650	10	0.10%
Biomedical Engineers	15,220	16,590	1,370	2.90%
Chemical Engineers	30,970	27,860	-3,110	-3.50%
Civil Engineers	261,360	254,130	-7,230	-0.90%
Computer Hardware Engineers	73,370	71,990	-1,380	-0.60%
Electrical Engineers	154,670	154,250	-420	-0.10%
Electronics Engineers, except Computer	139,930	136,310	-3,620	-0.90%
Environmental Engineers	52,590	50,350	-2,240	-1.40%
Health and Safety Engineers, except Mining Safety Engineers and Inspectors	25,190	23,170	-2,020	-2.70%
Industrial Engineers	214,580	211,490	-3,090	-0.50%
Marine Engineers and Naval Architects	6,480	5,470	-1,010	-5.50%
Materials Engineers	24,160	22,160	-2,000	-2.80%
Mechanical Engineers	233,610	238,260	4,650	0.70%
Mining and Geological Engineers, incl. Mining Safety Engineers	6,900	6,630	-270	-1.30%
Nuclear Engineers	16,640	18,430	1,790	3.50%
Petroleum Engineers	20,880	30,880	10,000	13.90%
Engineers, All Other	169,240	125,590	-43,650	-9.50%
Life Scientists				
Animal Scientists	2,760	2,190	-570	-7.40%
Food Scientists and Technologists	10,510	12,040	1,530	4.60%
Soil and Plant Scientists	10,790	11,860	1,070	3.20%
Biochemists and Biophysicists	22,230	25,160	2,930	4.20%
Microbiologists	15,750	17,660	1,910	3.90%
Zoologists and Wildlife Biologists	17,780	18,380	600	1.10%

Occupational Group	Employment, 2008	Employment, 2011	Employment Change, Number	Employment Change, Compound Annual Growth Rate
Biological Scientists, All Other	28,290	31,550	3,260	3.70%
Conservation Scientists	15,830	18,910	3,080	6.10%
Foresters	10,160	9,000	-1,160	-4.00%
Epidemiologists	4,370	4,610	240	1.80%
Medical Scientists, except Epidemiologists	99,750	95,220	-4,530	-1.50%
Life Scientists, All Other	12,030	10,020	-2,010	-5.90%
Physical Scientists				
Astronomers	1,280	2,080	800	17.60%
Physicists	14,810	16,220	1,410	3.10%
Atmospheric and Space Scientists	8,860	9,640	780	2.90%
Chemists	83,080	80,040	-3,040	-1.20%
Materials Scientists	9,650	7,900	-1,750	-6.50%
Environmental Scientists and Specialists, including Health	80,120	83,090	2,970	1.20%
Geoscientists, except Hydrologists and Geographers	31,260	32,490	1,230	1.30%
Hydrologists	7,590	6,960	-630	-2.80%
Physical Scientists, All Other	22,900	25,790	2,890	4.00%

Source: CRS analysis of Occupational Employment Survey data, 2008-2011, Bureau of Labor Statistics, U.S. Department of Labor, http://www.bls.gov/oes/tables.htm.

The physical sciences occupations with the largest growth were environmental scientists and specialists (2,970, 1.2% CAGR); physical scientists, all other (2,890, 4.0% CAGR); physicists (1,410, 3.1% CAGR); and geoscientists (1,230, 1.3%). Employment in three physical sciences occupations declined: chemists (-3,040, -1.2% CAGR), material scientists (-1,750, 6.5% CAGR), and hydrologists (-630, -2.8% CAGR).

Employment grew in each of the S&E managers occupations. The largest growth was in computer and information systems managers (24,010, 2.8%). Natural science managers had the fastest growth rate (4,450, 3.3%).

Architectural and engineering managers grew by 2,230, with a growth rate of 0.4%.

Table 5 shows the 10 S&E occupations with the largest employment growth. The top five occupations are computer occupations, followed by aerospace engineers and petroleum engineers.

Table 5. S&E Occupations with the Largest Employment Growth, 2008-2011

Rank	S&E Occupation	Employment Growth
1	Computer Support Specialists	86,970
2	Software Developers, Applications	45,720
3	Information Security Analysts, Web Developers, and Computer Network Architects	42,260
4	Computer and Information Systems Managers	24,010
5	Network and Computer Systems Administrators	13,950
6	Aerospace Engineers	11,600
7	Petroleum Engineers	10,000
8	Software Developers, Systems Software	5,220
9	Mechanical Engineers	4,650
10	Natural Sciences Managers	4,450

Source: CRS analysis of Occupational Employment Survey data, 2008-2011, BLS, U.S. Department of Labor.

Table 6. S&E Occupations with the Largest Employment Losses, 2008-2011

Rank	S&E Occupation	Employment Growth
1	Computer Programmers	-74,130
2	Engineers, All Other	-43,650
3	Computer Occupations, All Other	-14,150
4	Database Administrators	-7,270
5	Civil Engineers	-7,230
6	Mathematical Science Occupations, All Other	-5,340
7	Medical Scientists, except Epidemiologists	-4,530
8	Electronics Engineers, except Computer	-3,620
9	Chemical Engineers	-3,110
10	Industrial Engineers	-3,090

Source: CRS analysis of Occupational Employment Survey data, 2008-2011, Bureau of Labor Statistics, U.S. Department of Labor, http://www.bls.gov/oes/tables.htm.

Table 7. S&E Occupations with the Fastest Growth Rates, 2008-2011

Rank	S&E Occupation	Employment Growth Rate
1	Astronomers	17.60%
2	Petroleum Engineers	13.90%
3	Conservation Scientists	6.10%
4	Information Security Analysts, Web Developers, and Computer Network Architects	5.80%
5	Aerospace Engineers	5.40%
6	Computer Support Specialists	5.10%
7	Statisticians	4.80%
8	Food Scientists and Technologists	4.60%
9	Biochemists and Biophysicists	4.20%
10	Physical Scientists, All Other	4.00%

Source: CRS analysis of Occupational Employment Survey data, 2008-2011, Bureau of Labor Statistics, U.S. Department of Labor, http://www.bls.gov/oes/tables.htm.

Table 6 shows the 10 S&E occupations with the largest employment losses. The occupation with the greatest employment loss is computer programmers. Some have speculated that some of the losses in computer programmers may be due to reclassification of these positions as other computer occupations (e.g., software developers).

Table 7 shows the 10 S&E occupations with the fastest growth rates. The occupation with the fastest growth rate was astronomers, though the growth in the number of jobs (800) was small compared to some S&E occupations. In contrast, petroleum engineers, the second fastest growing S&E occupation, was also the seventh ranked occupation in terms of job growth (10,000).

Table 8 shows the 10 S&E occupations with the slowest growth rates. This group includes at least one occupation from each of the engineering, physical sciences, life sciences, mathematics, and computer occupational groups.

Wage Trends

Between 2008 and 2011, mean wages for each S&E occupational group grew at slower compound annual growth rates than the overall mean wage growth rate for all occupations, and only somewhat faster than inflation. **Figure 4** illustrates the nominal and inflation-adjusted compound annual growth rates for each S&E occupational group, as well as for all occupations. The nominal growth rate of mean wages for all occupations during this period was 2.3%, while the fastest growth rate in the S&E occupational groups was for physical scientists (2.2%) and mathematical occupations (2.2%), followed by engineers (2.1%), S&E managers (2.0%), computer occupations (1.8%), and life scientists (1.5%). Life scientists experienced nearly no real (inflation-adjusted) growth in mean wages between 2008 and 2011, while the other S&E occupational categories grew by less than 1% in inflation-adjusted terms, below the overall growth rate for all occupations.

Unemployment Trends

Figure 5 illustrates unemployment rates for the S&E occupational groups, as well as all for all workers (16 years and over) and selected professional and related occupations (e.g., lawyers, dentists, physicians and surgeons, registered nurses) for the years 2008-2011. This figure provides a perspective on how the unemployment rates of S&E occupational groups compare to the overall unemployment rate and other selected professional and related occupations, as well as how these rates changed during this period.

Compared to the overall workforce, S&E occupational groups had significantly lower unemployment rates for the 2008-2011 period. In general, though, the professional occupations (of which the S&E occupations are a part) historically have had a lower unemployment rate than the workforce as a whole.[22] As shown in **Figure 5**, the S&E occupational groups generally had unemployment rates that were comparable or higher than the rates for selected professional occupations during the 2008-2011 period. Also, the unemployment rates for the S&E occupational groups grew more than did the unemployment rates for the other selected professional occupations included in **Figure 5** (e.g., lawyers, dentists, physicians and surgeons, registered nurses). Consequently, the separation between the unemployment rates of the S&E occupational groups and the other selected professional occupations grew during this period. In 2008, the unemployment rates for each S&E occupational group and each selected professional occupation were below 3%. In 2011, the unemployment rates for each of the selected professional occupations remained below 3% while the unemployment rates for each of the S&E occupational groups were above 3%.

Employment Projections, 2010-2020

This section provides an analysis of the Bureau of Labor Statistics occupational employment projections data for the 2010-2020 period.

Table 8. S&E Occupations with the Slowest Growth Rates, 2008-2011

Rank	S&E Occupation	Employment Growth Rate
1	Mathematical Science Occupations, All Other	-42.40%
2	Engineers, All Other	-9.50%
3	Animal Scientists	-7.40%
4	Computer Programmers	-6.70%
5	Materials Scientists	-6.50%
6	Life Scientists, All Other	-5.90%
7	Marine Engineers and Naval Architects	-5.50%
8	Foresters	-4.00%
9	Chemical Engineers	-3.50%
10	Materials Engineers	-2.80%

Source: CRS analysis of Occupational Employment Survey data, 2008-2011, Bureau of Labor Statistics, U.S. Department of Labor, http://www.bls.gov/oes/tables.htm

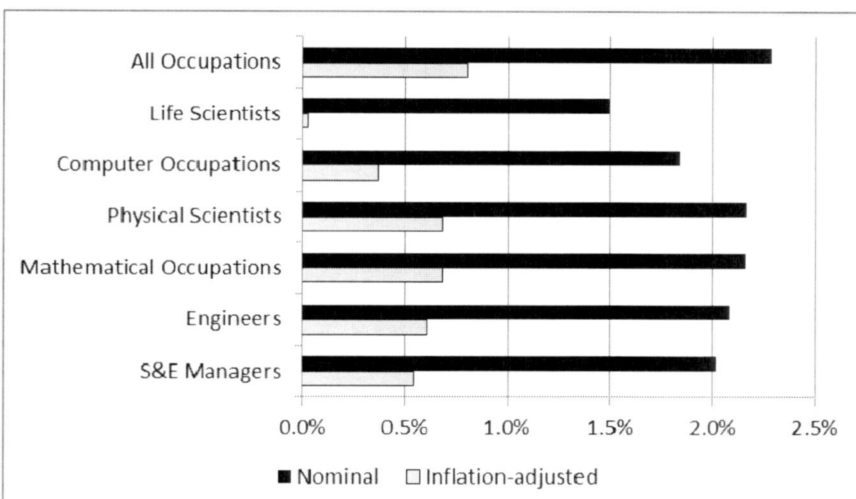

Source: CRS analysis of Occupational Employment Survey data, 2008-2011, http://www.bls.gov/oes/tables.htm, using Bureau of Labor Statistics Consumer Price Index Inflation Calculator, http://data.bls.gov/cgi-bin/cpicalc.pl.

Figure 4. Nominal and Inflation-adjusted Compound Annual Growth Rates of Mean Wages in S&E Occupational Groups, 2008-2011.

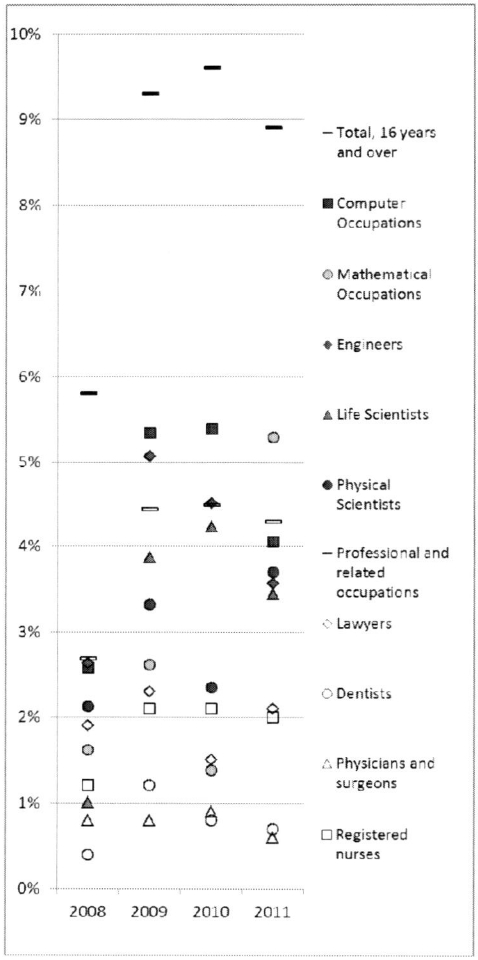

Source: CRS analysis of unpublished 2011data from the Current Population Survey, BLS.

Figure 5. Unemployment Rates for S&E Occupational Groups, the Overall Workforce, and Other Selected Professional and Related Occupations, 2008-2011.

Scientists and Engineers in Aggregate

An analysis of Bureau of Labor Statistics employment projections indicates that the science and engineering workforce will grow by 1.1 million (18.0%) jobs between 2010 and 2020, a CAGR of 1.7%. This growth rate is somewhat higher than the growth rate projected for all occupations (1.3%) during this period.[23]

In addition to the job openings created by growth in the number of jobs in S&E occupations, BLS projects that an additional 1.3 million scientists and engineers will be needed to replace those who are expected to exit the S&E occupations during this period due to retirement, death, career change, etc. (i.e., net replacements). BLS projects a total of 2.4 million job openings in S&E occupations due to growth and net replacements during this period.

Science and Engineering Occupational Groups

Projection data for science and engineering occupational groups is provided in **Table 9**, which includes the following data for each group: 2010 actual employment, 2020 projected employment, the change in the number of jobs between 2010 and 2020, the total percentage increase in the number of jobs, the compound annual growth rate in the number of jobs, and the total job openings due to growth and net replacements.

Among the S&E occupational groups, computer occupations is projected to see the fastest employment growth (2.0% CAGR), the largest increase in the number employed (758,900), and the largest number of job openings (1,336,400). Computer occupations, which accounted for 57.5% of all S&E jobs in 2010, are projected to account for 68.5% of the total growth in S&E occupations between 2010 and 2020. (See **Figure 6**.) As a result, the share of all S&E jobs accounted for by computer occupations is projected to rise from 55.6% in 2010 to 57.5% in 2020.

Life scientists is the only other S&E occupational group projected to account for a greater share (5.3%) of total S&E job growth than its share of total 2010 S&E employment (4.6%). As a result, life scientists' share of S&E employment is projected to increase slightly to 4.7% in 2020. The mathematical occupations group is projected to account for 1.8% of S&E job growth during this period, approximately the same as its 1.9% share of S&E occupational employment in 2010.

The occupational groups that are projected to have growth slower than their share of total 2010 S&E employment are:

- **engineers**—projected to account for 14.5% of total S&E job growth during the 2010-2020 period, below its 24.6% share of S&E employment in 2010, thus reducing its projected share of 2020 S&E employment to 23.1%;

- **physical scientists**—projected to account for 3.2% of total S&E job growth during the 2010-2020 period, below its 4.6% share of S&E employment in 2010, thus reducing its projected share of 2020 S&E employment to 4.4%; and
- **S&E managers**—projected to account for 6.8% of total S&E job growth during the 2010-2020 period, below its 8.7% share of S&E employment in 2010, thus reducing its projected share of 2020 S&E employment to 8.4%.

Table 9. 2010-2020 Employment Projections for S&E Occupational Groups Numbers in thousands, except percent and CAGR

	Employment		Change, 2010-2020			Job Opening Due to Growth and Net Replacements
Occupations	**2010**	**2020**	**Number**	**%**	**CAGR**	
Computer Occupations	3,425.9	4,184.6	758.9	22.1	2.0	1,366.4
Mathematical Occupations	116.9	136.4	19.5	16.7	1.6	71.6
Engineers	1,519.0	1,679.6	160.3	10.6	1.0	525.9
Life Scientists	286.0	344.3	58.2	20.4	1.9	106.1
Physical Scientists	282.0	317.6	35.7	12.6	1.2	121.8
S&E Managers	534.0	608.8	74.8	14.0	1.3	186.0
S&E Occupations, Total	**6,163.8**	**7,271.3**	**1,107.4**	**18.0**	**1.7**	**2,377.8**
Other Selected Occupations						
Lawyers	728.2	801.8	73.6	10.1	1.0	212
Dentists	155.7	187.9	32.2	20.7	1.9	78.4
Physicians and Surgeons	691.0	859.3	168.3	24.4	2.2	305.1
Registered Nurses	2,737.4	3,449.3	711.9	26.0	2.3	1,207.4
Post-Secondary Teachers	1,756.0	2061.7	305.7	17.4	1.6	586.1
Total, All Occupations	**143,068.2**	**163,537.1**	**20,468.9**	**14.3**	**1.3**	**54,787.4**

Source: CRS analysis of Employment Projections, 2010-2020, Bureau of Labor Statistics, U.S. Department of Labor.

Notes: Numbers for S&E occupational groups may not add due to rounding of component occupations.

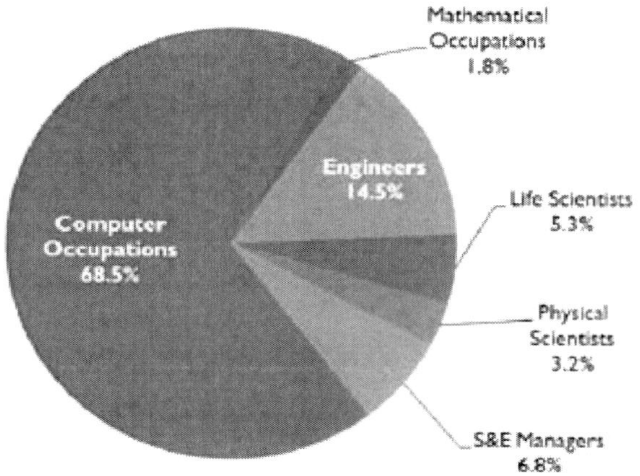

Source: CRS analysis of Employment Projections, 2010-2020, Bureau of Labor Statistics, U.S. Department of Labor.

Notes: Numbers are rounded and may not add to 100%.

Figure 6. Share of Total Projected S&E Occupational Job Growth, 2010-2020, by S&E Occupational Group.

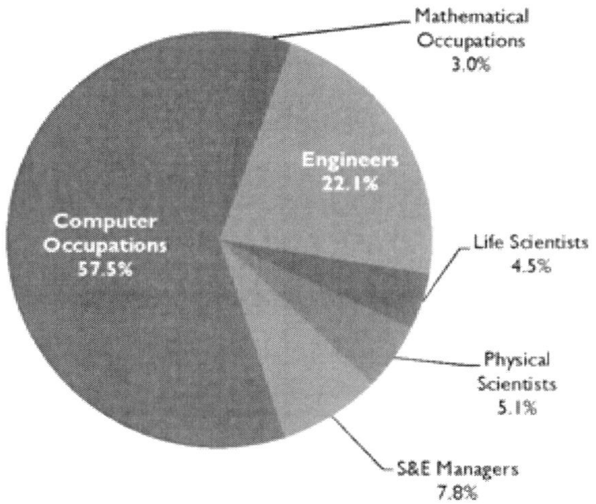

Source: CRS analysis of Employment Projections, 2010-2020, Bureau of Labor Statistics, U.S. Department of Labor.

Figure 7. Share of Total Projected S&E Occupational Job Openings (Job Growth plus Net Replacement Needs), 2010-2020, by S&E Occupational Group.

For some S&E occupational groups, the number of openings resulting from growth in the number employed is comparable to the number of openings resulting from those exiting the occupations (net replacements). For example, BLS projects an increase of 758,900 jobs in the computer occupations between 2010 and 2020. During the same period, BLS projects that there will be 607,500 job openings in computer occupations due to net replacement needs. For other S&E occupational groups, net replacement needs greatly exceed the number of projected new jobs in the occupation. For example, BLS projects 365,600 job openings due to net replacement needs in the engineering occupations between 2010 and 2020, and job growth of 160,300. **Figure 7** illustrates the composition of total projected S&E job openings (due to growth and net replacements) by S&E occupational group.

Table 10. S&E Occupations with the Highest Projected Growth in Jobs and Other Selected Occupations, 2010-2020

Rank	S&E Occupation	Projected Average Annual Job Growth[a]
1	Software Developers, Applications	14,380
2	Software Developers, Systems Software	12,720
3	Computer Systems Analysts	12,040
4	Computer Support Specialists	11,000
5	Network and Computer Systems Administrators	9,660
6	Information Security Analysts, Web Developers, and Computer Network Architects	6,570
7	Computer and Information Systems Managers	5,580
8	Civil Engineers	5,110
9	Computer Programmers	4,370
10	Medical Scientists, except Epidemiologists	3,640
Other Selected Occupations with High Projected Growth		
	Registered Nurses	71,190
	Retail Salespersons	70,680
	Home Health Aides	70,630
	Personal Care Aides	60,700
	Office Clerks, General	48,950

Source: CRS analysis of Employment Projections, 2010-2020, Bureau of Labor Statistics, U.S. Department of Labor.

[a] The numbers in this column are derived by dividing the net job creation during the 2010-2020 projection period for each occupation by 10 to get the average annual number of net new jobs created.

Detailed Science and Engineering Occupations

The Bureau of Labor Statistics' projected job growth and projected total job openings (job growth plus net replacements) for the S&E occupations vary substantially during the 2010-2020 projection period. Tables 12-17 show the top ten S&E occupations in terms of job growth, job losses, and job openings.

Table 10 shows the ten S&E occupations with the highest projected growth in jobs. Seven of the 10 S&E occupations on this list are in the computer occupations. One of the remaining three occupations is computer and information systems managers. The only non-IT occupations in the top ten are civil engineers and medical scientists.

Table 11 shows the ten S&E occupations with the smallest projected growth in jobs. The list includes occupations from physical and life sciences, mathematics, and engineering occupations. The number of new jobs projected to be created in these ten occupations total an average of less than 600 per year.

Table 11. S&E Occupations with the Smallest Projected Growth in Jobs, 2010-2020

Rank	S&E Occupation	Projected Average Annual Job Growth[a]
1	Agricultural Engineers	20
2	Astronomers	20
3	Mathematical Science Occupations, All Other	30
4	Animal Scientists	40
5	Mathematicians	50
6	Foresters	50
7	Mining and Geological Engineers, incl. Mining Safety Engineers	60
8	Materials Scientists	90
9	Marine Engineers and Naval Architects	100
10	Atmospheric and Space Scientists	100

Source: CRS analysis of Employment Projections, 2010-2020, Bureau of Labor Statistics, U.S. Department of Labor.

[a.] The numbers in this column are derived by dividing the net job creation during the 2010-2020 projection period for each occupation by 10 to get the average annual number of net new jobs created.

Table 12. S&E Occupations with the Fastest Projected Job Growth, 2010-2020

Rank	S&E Occupation	Projected Job Growth Rate (CAGR)	Projected Average Annual Job Growth[a]
1	Biomedical Engineers	4.9%	970
2	Medical Scientists, except Epidemiologists	3.2%	3,640
3	Software Developers, Systems Software	2.8%	12,720
4	Biochemists and Biophysicists	2.7%	770
5	Database Administrators	2.7%	3,390
6	Network and Computer Systems Administrators	2.5%	9,660
7	Software Developers, Applications	2.5%	14,380
8	Actuaries	2.4%	580
9	Epidemiologists	2.0%	120
10	Computer Systems Analysts	2.0%	12,040
Rank	**S&E Occupation**	**Projected Job Growth Rate (CAGR)**	**Projected Average Annual Job Growth[a]**
Other Occupations with the Fastest Projected Growth			
	Personal Care Aides	5.5%	60,700
	Home Health Aides	5.4%	70,630
	Helpers, Brickmasons, Blockmasons, Stonemasons, and Tile and Marble Setters	4.9%	1,760
	Helpers, Carpenters	4.5%	2,590
	Veterinary technologists and technicians	4.3%	4,170
All Occupations		1.3%	2,046,890

Source: CRS analysis of Employment Projections, 2010-2020, Bureau of Labor Statistics, U.S. Department of Labor.

[a] The numbers in this column are derived by dividing the net job creation during the 2010-2020 projection period for each occupation by 10 to get the average annual number of net new jobs created.

Table 12 shows the 10 S&E occupations with the fastest projected job growth. Biomedical engineers (4.9% CAGR) and medical scientists (3.2% CAGR) are the fastest growing S&E occupations. The remaining occupations on the list range from 2.0%-2.8% CAGR, faster than the overall projected job growth rate for all occupations (1.3% CAGR). The list includes five computer occupations.

Table 13. S&E Occupations with the Slowest Projected Job Growth, 2010-2020

Rank	S&E Occupation	Projected Job Growth Rate (CAGR)	Projected Average Annual Job Growth[a]
1	Chemists	0.4%	320
2	Foresters	0.4%	50
3	Aerospace Engineers	0.5%	400
4	Electronics Engineers, except Computer	0.5%	680
5	Conservation Scientists	0.5%	120
6	Computer Occupations, All Other	0.6%	1,230
7	Chemical Engineers	0.6%	180
8	Biological Scientists, All Other	0.6%	220
9	Industrial Engineers	0.6%	1,310
10	Engineers, All Other	0.6%	1,030

Source: CRS analysis of Employment Projections, 2010-2020, Bureau of Labor Statistics, U.S. Department of Labor.

a. The numbers in this column are derived by dividing the net job creation during the 2010-2020 projection period for each occupation by 10 to get the average annual number of net new jobs created.

Table 13 shows the 10 S&E occupations with the slowest projected job growth, ranging between 0.4%-0.6% CAGR, well below the overall projected job growth rate of 1.3% CAGR. The list includes five engineering occupations.

Table 14 shows the 10 S&E occupations with the most projected job openings (jobs growth plus net replacements). This category shows where the most job opportunities are projected to be. The top seven occupations are computer occupations, with computer and information systems managers in the ninth spot. The other two occupations on the list are engineering occupations, civil engineering and mechanical engineering.

Table 15 shows the 10 S&E occupations with the fewest projected job openings. This list includes occupations from the life sciences, engineering, physical sciences, and mathematics occupations.

Table 14. S&E Occupations with the Most Projected Job Openings, 2010-2020

Rank	S&E Occupation	Projected Average Annual Job Growth[a]
1	Computer Support Specialists	26,950
2	Computer Systems Analysts	22,250
3	Software Developers, Applications	19,790
4	Software Developers, Systems Software	16,800
5	Network and Computer Systems Administrators	15,530
6	Computer Programmers	12,800
7	Information Security Analysts, Web Developers, and	11,030
8	Civil Engineers	10,440
9	Computer and Information Systems Managers	10,280
10	Mechanical Engineers	9,960

Source: CRS analysis of Employment Projections, 2010-2020, Bureau of Labor Statistics, U.S. Department of Labor.

[a.] The numbers in this column are derived by dividing the net job creation during the 2010-2020 projection period for each occupation by 10 to get the average annual number of net new jobs created.

Table 15. S&E Occupations with the Fewest Projected Job Openings, 2010-2020

Rank	S&E Occupation	Projected Average Annual Job Growth[a]
1	Agricultural Engineers	80
2	Astronomers	90
3	Epidemiologists	150
4	Miscellaneous Mathematical Science Occupations	170
5	Animal Scientists	180
6	Foresters	190
7	Mining and Geological Engineers, incl. Mining Safety Engineers	200
8	Life Scientists, All Other	210
9	Atmospheric and Space Scientists	210
10	Marine Engineers and Naval Architects	230

Source: CRS analysis of Employment Projections, 2010-2020, Bureau of Labor Statistics, U.S. Department of Labor.

CONCLUSION

The adequacy of the U.S. science and engineering workforce has been an ongoing concern of Congress for more than 60 years. Scientists and engineers are widely believed to be essential to U.S. technological leadership, innovation, manufacturing, and services, and thus vital to U.S. economic strength, national defense, and other societal needs. Congress has enacted many programs to support the education and development of scientists and engineers. Congress has also undertaken broad efforts improve science, technology, engineering, and math (STEM) skills to prepare a greater number of students to pursue science and engineering (S&E) degrees. Some policymakers have sought to increase the number of foreign scientists and engineers working in the United States through changes in visa and immigration policies.

Scientists and engineers are widely believed to be essential to U.S. technological leadership, innovation, manufacturing, and services, and thus vital to U.S. economic strength, national defense, and other societal needs (e.g., treating and preventing diseases, ensuring access to affordable energy, protecting and restoring the environment). However, there are varying perspectives with respect to the question of the existence of a shortage of scientists and engineers in the United States, what the nature of such a shortage might be (e.g., too few people with S&E degrees, a mismatch of worker skills and employer needs), and whether the federal government should undertake policy interventions to address a putative shortage or allow market forces to work in this labor market.

Many policymakers, business leaders, academicians, S&E professional society analysts, economists, and others hold diverse views with respect to the adequacy of the S&E workforce and related policy issues. Here are some general characterizations of those views:

- **There is a shortage.** There is a shortage (or a looming shortage) of scientists and engineers (or, alternatively, an inadequate supply of workers with degrees in science and engineering fields), potentially resulting in the loss of U.S. scientific, engineering, technological, and industrial leadership and consequent effects on areas such as economic growth, job creation, standard of living, and national security.[24]
- **There is not a shortage.** Assertions of a broad shortage of scientists and engineers are not supported by the data when considering

indicators such as employment growth, wage growth, and unemployment rates.[25]

- **More scientists and engineers are needed regardless of the existence of a shortage.** Regardless of whether demand currently exceeds supply, increasing the number of U.S. scientists and engineers will increase U.S. innovation, economic performance, and job creation. Even if there is not a shortage of scientists and engineers, jobs in many occupations require a higher level of STEM knowledge than ever before.[26] Students who earn S&E degrees gain thinking skills, problem-solving skills, and STEM knowledge that will enable them to be successful not only in S&E occupations, but also in S&E-related careers or to apply their S&E knowledge and skills in non-S&E fields.[27] Historically, federal policies, programs, and investments have contributed to the development of the United States' scientific and engineering workforce.
- **Government interventions in the S&E labor market to address perceived shortages may introduce inefficiencies.** Federal government efforts to increase the number of scientists and engineers by incentivizing the pursuit of degrees in S&E disciplines and/or increasing immigration quotas may result in less efficient operation of the S&E labor market (e.g., too many students educated in S&E for the number of jobs available; lower salaries for graduates who find S&E jobs).[28]
- **Workforce projections are unreliable for predicting shortages.** Long-term projections for S&E occupations are unreliable. [29] Relying on such projections could possibly result in the preparation of too many or too few students with S&E degrees or in mismatches between the students' education and market needs. Among the difficulties in making long term projections are unexpected changes in: the mix of industrial output or employment due to technological or market changes; the use of substitution (e.g., capital for labor) resulting from changes in factor prices; retirement behavior; the availability of foreign labor sources; labor market demographics; and government policies.[30]
- **There may be shortages in certain industries, occupations, or fields.** Shortages may exist in some S&E occupations or for certain employers, for example in new and emerging S&E fields (e.g., nanotechnology), cyclical industries (e.g., aerospace); in fields where foreign scientists and engineers may not be employed due to export

control laws; and for employers otherwise limited, in general or for specific purposes, to using only U.S. citizens.

  - **The labor market will resolve such needs.** If markets are allowed to operate freely (i.e., without government interventions), any short-term "shortages" will be resolved as wages equilibrate demand and supply, labor supply increases (i.e., as more students earn S&E degrees) in response to market signals, or through substitution of alternative inputs.[31]
  - **The potential adverse consequences of even discrete shortages require government interventions.** These shortages should be met with federal efforts to increase supply or the United States may face the loss of technological leadership in new and emerging fields, lower economic performance, and diminished national security.[32]

- **Industry assertions of shortages are driven by a desire to reduce costs, increase current knowledge.** Industry assertions of S&E shortages are driven primarily by a desire to lower their labor costs through increased supply and by providing a continuous stream of young, lower-cost recent college graduates (i.e., through education and training and by increases in immigrant visas) who can be hired to replace older, higher-cost workers with less current knowledge.[33]
- **The real issue is a skills mismatch, not a shortage of people.** The difficulty employers have in meeting their S&E workforce needs (in particular their information technology workforce needs) results primarily from a mismatch between the specific skills—or combinations of knowledge, skills, and experience—needed by employers and those held by S&E workers.[34]
- **Expanding immigration can help address the shortage.** Immigration policies directed at increasing the number of foreign scientists and engineers in the United States puts the creativity of the world's best and brightest to work for the U.S. economy and prevents the loss of U.S.-educated foreign nationals with S&E degrees (i.e., returning to their countries of origin, working in countries other than the United States).[35]
- **Expanding immigration will dampen the market signals that would otherwise drive more U.S. students into science and engineering.** Visa and immigration policies directed at increasing the number of foreign scientists and engineers in the United States may, by increasing the overall supply of scientists and engineers, depress

wages, increase unemployment, and reduce career opportunities for U.S. scientists and engineers; discourage American students from pursuing S&E degrees and careers; and cloud labor market signals (e.g., wage increases, lower unemployment) to students considering pursing S&E degrees and careers.[36]

- **U.S. students lag those of other nations in STEM knowledge; federal efforts to improve STEM education are needed.** U.S. students lag foreign students in STEM knowledge possibly resulting in fewer and/or less-talented U.S. scientists and engineers, lower economic growth, and reduced economic competitiveness.[37] Federal policies and programs can help to build a stronger K-12 STEM education system.
- **International assessments do not reflect the adequacy of U.S. student STEM knowledge.** Standardized tests used to compare the STEM knowledge of U.S. K- 12 students to those of other nations does not appropriately reflect the STEM knowledge of U.S. students, the adequacy of their preparation to pursue S&E degrees and occupations, or their future capabilities as scientists and engineers.[38]

These disparate perspectives contribute to a variety of opinions on the role(s) the federal government should play in fostering the development of the S&E workforce including the merits of federal policies focused on:

- increasing the number of students pursuing S&E degrees;
- increasing the number of foreign scientists and engineers admitted to the United States;
- increasing the number and share of underrepresented minorities and women in science and engineering;
- improving K-12 STEM education; and
- improving career information and counseling for high school students.

As Congress considers approaches to bolstering U.S. competitiveness and scientific, engineering, technological, and industrial leadership, it may wish to consider these perspectives and opinions.

APPENDIX. S&E OCCUPATIONAL DESCRIPTIONS AND ENTRY-LEVEL EDUCATION REQUIREMENTS

Occupation	Description	Entry-level Education
Computer Occupations		
Computer and Information Research Scientists	Computer and information research scientists invent and design new technology and find new uses for existing technology. They study and solve complex problems in computing for business, science, medicine, and other uses.	Doctoral or professional degree
Computer Programmers	Computer programmers write code to create software programs. They turn the program designs created by software developers and engineers into instructions that a computer can follow.	Bachelor's degree
Computer Support Specialists	Computer support specialists provide help and advice to people and organizations using computer software or equipment. Some, called	Some college, no degree
	technical support specialists, support information technology (IT) employees within their organization. Others, called help- desk technicians, assist non-IT users who are having computer problems.	
Computer Systems Analysts	Computer systems analysts study an organization's current computer systems and procedures and make recommendations to management to help the organization operate more efficiently and effectively. They bring business and information technology (IT) together by understanding the needs and limitations of both.	Bachelor's degree
Database Administrators	Database administrators use software to store and organize data, such as financial information and customer shipping records. They make sure that data are available to users and are secure from unauthorized access.	Bachelor's degree
Information Security Analysts, Web Developers, and Computer Network Architects	Information security analysts, web developers, and computer network architects all use information technology (IT) to advance their organization's goals. Security analysts ensure a firm's information stays safe from cyberattacks. Web developers create websites to help firms have a public face. Computer network architects create the internal networks all workers within organizations use.	Bachelor's degree

(Continued)

Occupation	Description	Entry-level Education
Network and Computer Systems Administrators	Network and computer systems administrators are responsible for the day-to-day operation of an organization's computer networks. They organize, install, and support an organization's computer systems, including local area networks (LANs), wide area networks (WANs), network segments, intranets, and other data communication systems.	Bachelor's degree
Software Developers	Software developers are the creative minds behind computer programs. Some develop the applications that allow people to do specific tasks on a computer or other device. Others develop the underlying systems that run the devices or control networks.	Bachelor's degree
Actuaries	Actuaries analyze the financial costs of risk and uncertainty. They use mathematics, statistics, and financial theory to assess the risk that an event will occur and to help businesses and clients develop policies that minimize the cost of that risk.	Bachelor's degree
Mathematicians	Mathematicians use high-level mathematics and technology to develop new mathematical principles, understand relationships between existing principles, and solve real-world problems.	Master's degree
Operations Research Analysts	Operations research analysts use advanced methods of analysis to help organizations solve problems and make better decisions.	Bachelor's degree
Statisticians	Statisticians use mathematical techniques to analyze and interpret data and draw conclusions.	Master's degree
Engineers		
Aerospace Engineers	Aerospace engineers design aircraft, spacecraft, satellites, and missiles. In addition, they test prototypes to make sure that they function according to design.	Bachelor's degree
Agricultural Engineers	Agricultural engineers—also known as biological and agricultural engineers—work on a variety of activities. These activities range from aquaculture (raising food, such as fish, that thrive in water) to land farming to forestry; from developing biofuels to improving conservation; from planning animal environments to finding better ways to process food.	Bachelor's degree

Occupation	Description	Entry-level Education
Biomedical Engineers	Biomedical engineers analyze and design solutions to problems in biology and medicine, with the goal of improving the quality and effectiveness of patient care.	Bachelor's degree
Chemical Engineers	Chemical engineers apply the principles of chemistry, biology, and physics to solve problems. These problems involve the production or use of chemicals, fuel, drugs, food, and many other products. They design processes and equipment for large-scale safe and sustainable manufacturing, plan and test methods of manufacturing products and treating byproducts, and supervise production.	Bachelor's degree
Civil Engineers	Civil engineers design and supervise large construction projects, including roads, buildings, airports, tunnels, dams, bridges, and systems for water supply and sewage treatment.	Bachelor's degree
Computer Hardware Engineers	Computer hardware engineers research, design, develop, and test computer equipment such as chips, circuit boards, or routers. By solving complex problems in computer hardware, these engineers create rapid advances in computer technology.	Bachelor's degree
Electrical and Electronics Engineers	Electrical engineers design, develop, test, and supervise the manufacturing of electrical equipment such as electric motors, radar and navigation systems, communications systems, and power generation equipment. Electronics engineers design and develop electronic equipment, such as broadcast and communications systems—from portable music players to global positioning systems (GPS).	Bachelor's degree
Environmental Engineers	Environmental engineers use the principles of engineering, soil science, biology, and chemistry to develop solutions to environmental problems. They are involved in efforts to improve recycling, waste disposal, public health, and control of water and air pollution.	Bachelor's degree
Health and Safety Engineers	Health and safety engineers develop procedures and design systems to keep people from getting sick or injured and to keep property from being damaged. They combine knowledge of health or safety and of systems engineering to make sure that chemicals, machinery, software, furniture, and	Bachelor's degree

(Continued)

Occupation	Description	Entry-level Education
	other products are not going to cause harm to people or buildings.	
Industrial Engineers	Industrial engineers find ways to eliminate wastefulness in production processes. They devise efficient ways to use workers, machines, materials, information, and energy to make a product or provide a service.	Bachelor's degree
Marine Engineers and Naval Architects	Marine engineers and naval architects design, build, and maintain ships from aircraft carriers to submarines, from sailboats to tankers. Marine engineers work on the mechanical systems, such as propulsion and steering. Naval architects work on the basic design, including the form and stability of hulls.	Bachelor's degree
Materials Engineers	Materials engineers develop, process, and test materials used to create a range of products, from computer chips and aircraft wings to golf clubs and snow skis. They also help select materials and develop new ways to use materials.	Bachelor's degree
Mechanical Engineers	Mechanical engineering is one of the broadest engineering disciplines. Mechanical engineers design, develop, build, and test mechanical devices, including tools, engines, and machines.	Bachelor's degree
Mining and Geological Engineers	Mining and geological engineers design mines for the safe and efficient removal of minerals, such as coal and metals, for manufacturing and utilities.	Bachelor's degree
Nuclear Engineers	Nuclear engineers research and develop the processes, instruments, and systems used to get benefits from nuclear energy and radiation. Many of these engineers find industrial and medical uses for radioactive materials—for example, in equipment used in medical diagnosis and treatment.	Bachelor's degree
Petroleum Engineers	Petroleum engineers design and develop methods for extracting oil and gas from deposits below the earth's surface. Petroleum engineers also find new ways to extract oil and gas from older wells.	Bachelor's degree
Life Scientists		
Agricultural and Food Scientists	Agricultural and food scientists work to ensure agricultural productivity and food safety.	Agricultural and food scientists: at

Occupation	Description	Entry-level Education
		least a bachelor's degree. Food scientists and soil and plant scientists: typically a bachelor's degree. Most animal scientists earn a Ph.D.
Biochemists and Biophysicists	Biochemists and biophysicists study the chemical and physical principles of living things and of biological processes such as cell development, growth, and heredity.	Doctoral or professional degree
Microbiologists	Microbiologists study the growth, development, and other characteristics of microscopic organisms such as bacteria, algae, and fungi.	Bachelor's degree
Zoologists and Wildlife Biologists	Zoologists and wildlife biologists study the characteristics and habitats of animals and wildlife.	Bachelor's degree
Conservation Scientists and Foresters	Conservation scientists and foresters manage overall land quality of forests, parks, rangelands, and other natural resources.	Bachelor's degree
Epidemiologists	Epidemiologists investigate the causes of disease and other public health problems to prevent them from spreading or from happening again. They report their findings to public policy officials and to the general public.	Master's degree
Medical Scientists	Medical scientists conduct research aimed at improving overall human health. They often use clinical trials and other investigative methods to reach their findings.	Doctoral or professional degree
Physical Scientists		
Physicists and Astronomers	Physicists and astronomers study the fundamental nature of the universe, ranging from the vastness of space to the smallest of subatomic particles. They develop new technologies, methods, and theories based on the results of their research that deepen our understanding of how things work and contribute to innovative, real-world applications.	Doctoral or professional degree

(Continued)

Occupation	Description	Entry-level Education
Atmospheric Scientists, Including Meteorologists	Atmospheric scientists study weather, climate, and other aspects of the atmosphere. They develop reports and forecasts from their analysis of weather and climate data.	Bachelor's degree
Chemists and Materials Scientists	Chemists and materials scientists study the structures, compositions, reactions, and other properties of substances. They use their knowledge to develop new and improved products, processes, and materials.	Bachelor's degree
Environmental Scientists and Specialists	Environmental scientists and specialists use their knowledge of the natural sciences to protect the environment. They identify problems and find solutions that minimize hazards to the health of the environment and the population.	Bachelor's degree
Geoscientists	Geoscientists study the physical aspects of the Earth, such as its composition, structure, and processes, to learn about its past, present, and future.	Bachelor's degree
Hydrologists	Hydrologists study water and the water cycle. They use their expertise to solve problems in the areas of water quality or availability.	Master's degree
S&E Managers		
Architectural and Engineering Managers	Architectural and engineering managers plan, coordinate, and direct activities in architecture and engineering, including research and development in these fields.	Bachelor's degree
Computer and Information Systems Managers	Computer and information systems managers, often called information technology managers (IT managers or IT project managers), plan, coordinate, and direct computer- related activities in an organization. They help determine the information technology goals of an organization and are responsible for implementing the appropriate computer systems to meet those goals.	Bachelor's degree
Natural Sciences Managers	Natural sciences managers supervise the work of scientists, including chemists, physicists, and biologists. They direct research and development projects and coordinate activities such as testing, quality control, and production.	Bachelor's degree

Source: *Occupational Outlook Handbook*, Bureau of Labor Statistics, Department of Labor, http://www.bls.gov/ ooh/management/home.htm.

End Notes

[1] For additional information about P.L. 110-69 and P.L. 111-358, see CRS Report R42430, *America COMPETES 2010 and the FY2013 Budget*, by Heather B. Gonzalez.

[2] For additional information, see CRS Report R42530, *Immigration of Foreign Nationals with Science, Technology, Engineering, and Mathematics (STEM) Degrees* , by Ruth Ellen Wasem.

[3] Among the key factors that labor economists examine for evidence of labor shortages are employment growth, wage growth, and unemployment rates relative to other occupations. See, for example, Carolyn M. Veneri, "Can Occupational Labor Shortages Be Identified Using Available Data?," *Monthly Labor Review*, March 1999, p. 18.

[4] For purposes of the BLS analysis, the authors defined the STEM occupation group as consisting of 97 occupations, including computer and math sciences, architecture and engineering, life and physical sciences, managerial and post- secondary teaching occupations associated with these functional areas, and two sales occupations that require scientific or technical education at the postsecondary level—sales engineers and wholesale and engineering manufacturing sales representatives of technical and scientific products. Ben Cover, John Jones, and Audrey Watson, "Science, Technology, Engineering, and Mathematics (STEM) Occupations:A Visual Essay," *Monthly Labor Review*, May 2011, p. 3.

[5] National Science Board, *Science and Engineering Indicators 2012*, January 2012, p. 3-10.

[6] National Science Board, *Science and Engineering Indicators 2008*, January 2008, p. 3-8.

[7] The Standard Occupational Classification system is a federal system that defines over 840 detailed occupations, and groups them into 461 broad occupations, 97 minor groups, and 23 major groups. Detailed occupations in the SOC with similar job duties, and in some cases skills, education, and/or training, are grouped together. The system is used by federal statistical agencies for the purpose of collecting, calculating, and disseminating data. First established in 1977, the SOC system has been revised periodically; the latest revision is the 2010 SOC.

[8] Occupational Employment Survey (OES) employment figures for the occupation group "architectural and engineering managers" are reported as a single number, thus the architectural managers are included in this group, though data on architect and other architectural-related occupations are not otherwise included in this report.

[9] Occupational Employment Survey, Bureau of Labor Statistics, U.S. Department of Labor, http://www.bls.gov/oes.

[10] Current Population Survey, Bureau of Labor Statistics, U.S. Department of Labor, http://www.bls.gov/cps.

[11] Employment Projections, Bureau of Labor Statistics, U.S. Department of Labor, http://www.bls.gov/emp.

[12] The Current Employment Statistics survey provides industry employment data used by BLS in making its biennial ten-year projections. The CES survey does not collect occupational information.

[13] Telephone conversation with Michael Wolf, economist, Division of Occupational Outlook, Office of Occupational Statistics and Employment Projections, Bureau of Labor Statistics, Department of Labor, March 4, 2013.

[14] BLS website, Occupational Employment Statistics, Frequently Asked Questions, http://www.bls.gov/oes/oes_ques.htm.

[15] BLS website, Employment Projections, Projections Methodology, http://www.bls.gov/emp/ep_projections_methods.htm.

[16] For links to past evaluations of BLS projections, see http://www.bls.gov/emp/ep_pub_projections_eval.htm. For the latest evaluation, see "Evaluating the 1996–2006

employment projections," by Ian D. Wyatt, *Monthly Labor Review*, September 2010, http://www.bls.gov/opub/mlr/2010/09/art3full.pdf.

[17] Science and engineering occupations are part of the larger category of "Professional and Related Occupations" used in the Current Population Survey.

[18] Mathematical occupations was the only S&E occupational group with a higher unemployment rate (5.3%) than the professional and related occupations group (4.3%).

[19] The occupational classification system used in the Current Population Survey is based on the 2010 Standard Occupational Classification System but differs somewhat from those used by in the Occupational Employment Survey.

[20] CPS does not publish unemployment rates for occupations with an employment base of less than 50,000.

[21] A compound annual growth rate (CAGR) is a calculated growth rate which, if applied year after year to a beginning amount reaches a specified final amount.

[22] For example, between 2000 and 2011, the unemployment rate for the overall workforce (age 16 years and over) was generally twice as high as that of professional and related occupations. In 2000, the unemployment rate for the overall workforce was 4.0%, while the unemployment rate for professional and related occupations was 1.9%; in 2011, the rates were 8.9% and 4.3%, respectively.

[23] CRS analysis of BLS 2010-2020 employment projections, http://www.bls.gov/emp.

[24] See, for example, National Research Council, *Rising Above the Gathering Storm: Energizing and Employing America for a Brighter Economic Future*, 2007, http://www.nap.edu/ catalog.php?record_id=11463; U.S. Department of Energy, *Secretary Chu, Intel President Discuss Need for More U.S. Engineers*, September 1, 2011, http://energy.gov/ articles/secretary-chu-intel-president-discuss-need-more-us-engineers; Shirley Ann Jackson, President, Rensselaer Polytechnic Institute, *The Quite Crisis: Falling Short in Producing American Scientific and Technical Talent,* Building Engineering and Science Talent (BEST), 2002; and Vinton G. Cerf, "How to Fire Up U.S. Innovation," *Wall Street Journal*, April 12, 2011, http://online.wsj.com/article/ SB10001424052748704461304576216911954533514.html; and Rodney C. Atkins, Senior Vice President, Systems and Technology Group, IBM, "America Desperately Needs More STEM Students. Here's How to Get Them," *Forbes*, July 9, 2012, http://www.forbes.com/ sites/forbesleadershipforum /2012/07/09/america-desperately-needs-more-stem-students-heres-how-to-get-them.

[25] See, for example, testimony of Ralph Gomory, President, Alfred P. Sloan Foundation, before the U.S. Congress, House Committee on Science and Technology, *The Globalization of R&D and Innovation, Part I*, 110th Cong., June 12, 2007 (Washington: GPO, 2008); testimony of Michael Teitelbaum, Vice President, Alfred P. Sloan Foundation and Harold Salzman, Senior Research Associate, The Urban Institute, before the U.S. Congress, House Committee on Science and Technology, *The Globalization of R&D and Innovation, Part IV*, 110th Cong., November 6, 2007 (Washington: GPO, 2008); Robert J. Samuelson, "Sputnik Scare, Updated" *Washington Post*, August 26, 2005, p. A27, http://www.washingtonpost. com/wp-dyn/content/article/2005/05/25/AR2005052501812. html; and Michael Teitelbaum, "The U.S. Science and Engineering Workforce: An Unconventional Portrait," *Pan-Organizational Summit on the U.S. Science and Engineering Workforce*, Government-Industry-University Research Roundtable, National Research Council, 2003, pp. 1-7, http://www.nap.edu/catalog.php?record_id=10727.

[26] See, for example, U.S. Congress Joint Economic Committee, Chairman's Staff, *STEM Education: Preparing for the Jobs of the Future*, April 2012, http://www.jec.senate.gov/ public/index.cfm?a=Files.Serve&File_id=6aaa7e1f-9586- 47be-82e7-326f47658320.

[27] See, for example, Vern Ehlers, before the U.S. Congress, House Committee on Science and Technology, *The Globalization of R&D and Innovation, Part IV*, 110th Cong., November 6, 2007 (Washington: GPO, 2008).

[28] See, for example, Leonard Lynn, Case Western Reserve University, and Hal Salzman, Rutgers University, "Dynamics of Engineering Labor Markets: Petroleum Engineering and

Responsive Supply," presentation at "U.S. Engineering in the Global Economy," sponsored by the Alfred P. Sloan Foundation, Cambridge, MA, September 26, 2011, http://policy.rutgers.edu/faculty/salzman/dynamics.pdf.

29 See, for example, Office of Technology Assessment, *Demographic Trends and the Scientific and Engineering Workforce*, OTA-TM-SET-35, December 1985, http://www.princeton.edu/~ota/disk2/1985/8507/8507.PDF.

30 See, for example, Richard B. Freeman, *Is a Great Labor Shortage Coming? Replacement Demand in the Global Economy*, National Bureau of Economic Research, Working Paper 12541, Cambridge, MA, September 2006, http://www.nber.org/papers/w12541.

31 See, for example, Richard B. Freeman, *Does Globalization of the Scientific/Engineering Workforce Threaten U.S. Economic Leadership?*, National Bureau of Economic Research, Working Paper 11457, Cambridge, MA, June 2005, http://www.nber.org/papers/w11457.pdf.

32 See, for example, National Research Council, *Rising Above the Gathering Storm: Energizing and Employing America for a Brighter Economic Future*, 2007.

33 See, for example, various writings of Norm Matloff, Professor of Computer Science, University of California at Davis, http://heather.cs.ucdavis.edu/matloff.html.

34 See, for example, Tom Kucharvy, *Solutions to STEM Skills Mismatch*, Beyond IT, February 25, 2012, http://beyond- it-inc.com/GKEblog/solutions-to-stem-skills-mismatch.html; and "Statistic of the Month: Investigating the Skills Mismatch," Center on International Education Benchmarking, July 31, 2012, http://www.ncee.org/2012/07/statistic-of- the-month-investigating-the-skills-mismatch.

35 See, for example, Vivek Wadhwa, Anna Lee Saxenian, Richard Freeman, and Alex Salever, *Losing the World's Best and Brightest: America's New Immigrant Entrepreneurs*, Ewing Marion Kauffman Foundation, March 2009, http://www.kauffman.org/uploadedFiles/ResearchAndPolicy/Losing_the_World%27s_Best_and_Brightest.pdf; The White House, "Fact Sheet: Fixing Our Broken Immigration System So Everyone Plays by the Rules," press release, January 29, 2013, http://www.whitehouse.gov/the-press-office/2013/01/29/fact-sheet-fixing-our-broken-immigration- system-so-everyone-plays-rules; and Robert D. Atkinson, *Eight Ideas for Improving the America COMPETES Act*, Information Technology and Innovation Foundation, March 2010, http://www.itif.org/files/2010-america-competes.pdf.

36 See, for example, Ross Eisenbrey, Vice President, Economic Policy Institute, "Op-Ed: America's Genius Glut," *New York Times*, February 7, 2013; Remarks of Brian Keane, Founder and CEO, Ameritas Technologies, and Neeraj Gupta, Founder and CEO, Systems in Motion, at Senate briefing on "Understanding the Impact of the H-1B Program: On the Economy, Employers and Workers," March 14, 2013, http://www.epi.org/files/2013/Keane_H- 1B_briefing_14_March_2013.pdf; and Stan Sorscher, Labor Represenative, Society of Professional Engineering Employees in Aerospace, *Flooding the STEM Labor Market*, March 3, 2013, http://www.ifpte.org/ downloads/issues/ 2013-3-3%20Flooding%20the%20STEM%20labor%20market.pdf.

37 U.S. Department of Education, "Secretary Arne Duncan's Remarks at OECD's Release of the Program for International Student Assessment (PISA) 2009 Results," press release, December 7, 2010, http://www.ed.gov/news/ speeches/secretary-arne-duncans-remarks-oecds-release-program-international-student-assessment; Paul E. Peterson, Ludger Woessmann, Eric A. Hanushek, and Carlos X. Lastra-Anadon, *Globally Challenged: Are U.S. Students Ready to Compete*, Harvard Kennedy School, Harvard University, PEPG Report No. 11-03, August 2011, http://www.hks.harvard.edu/pepg/PDF/Papers/PEPG11-03_GloballyChallenged.pdf; and Brandon Wright, "What do International Tests Really Show About U.S. Student Performance," Thomas B. Fordham Institute, January 24, 2013, http://www.edexcellence.net/commentary/education-gadfly-weekly/2013/january-24/what-do-international-tests-really- show-about-us-performance.html.

[38] Martin Carnoy and Richard Rothstein, *What Do International Tests Really Show About U.S. Student Performance*, Economic Policy Institute, January 15, 2013, http://www.epi.org/publication/us-student-performance-testing.

In: The U.S. Science and Engineering Workforce ISBN: 978-1-62948-932-2
Editors: Russell B. Hartley

Chapter 2

IMMIGRATION OF FOREIGN NATIONALS WITH SCIENCE, TECHNOLOGY, ENGINEERING, AND MATHEMATICS (STEM) DEGREES*

Ruth Ellen Wasem

SUMMARY

Although the United States remains the leading host country for international students in science, technology, engineering, or mathematics (STEM) fields, the global competition for talent has intensified. A record number of STEM graduates—both U.S. residents and foreign nationals—are entering the U.S. labor market, and there is a renewed focus on creating additional immigration pathways for foreign professional workers in STEM fields. Current law sets an annual worldwide level of 140,000 employment-based admissions, which includes the spouses and children in addition to the principal (i.e., qualifying) aliens. "STEM visa" is shorthand for an expedited immigration avenue that enables foreign nationals with graduate degrees in STEM fields to adjust to legal permanent resident (LPR) status without waiting in the queue of numerically limited LPR visas. The fundamental

* This is an edited, reformatted and augmented version of the Congressional Research Service Publication, CRS Report for Congress R42530, dated December 17, 2012.

policy question is should the United States create additional pathways for STEM graduates to remain in the United States permanently?

The number of full-time graduate students in science, engineering, and health fields who were foreign students (largely on F-1 nonimmigrant visas) grew from 91,150 in 1990 to 148,923 in 2009, with most of the increase occurring after 1999. Despite the rise in foreign student enrollment, the percentage of STEM graduate students with temporary visas in 2009 (32.7%) was comparable to 1990 (31.1%). Graduate enrollments in engineering fields have exhibited the most growth of the STEM fields in recent years. About 40,000 graduate degrees were awarded to foreign STEM students in 2009, with 10,000 of those going to Ph.D. recipients.

After completing their studies, foreign students on F-1 visas are permitted to participate in employment known as Optional Practical Training (OPT), which is temporary employment that is directly related to an F-1 student's major area of study. Generally, a foreign student may work up to 12 months in OPT status. In 2008, the Department of Homeland Security (DHS) expanded the OPT work period to 29 months for F-1 students in STEM fields.

Many F-1 visa holders (especially those who are engaged in OPT) often change their immigration status to become professional specialty workers (H-1B). Most H-1B beneficiaries are typically admitted to work in STEM occupations. In FY2010, the most recent year for which detailed data on H-1B beneficiaries (i.e., workers renewing their visas as well as newly arriving workers) are available, almost 91,000 H-1B workers were employed in computer-related occupations, and they made up 47% of all H-1B beneficiaries that year.

The H-1B visa and the OPT often provide the link for foreign students to become employment- based LPRs. In total, foreign nationals reporting STEM occupations made up 44% of all of the 676,642 LPRs who were employment-based principal immigrants during the decade of FY2000- FY2009. Of all of the LPRs reporting STEM occupations (297,668) over this decade, 52% entered as professional and skilled workers. STEM graduates seeking LPR status are likely to wait in line to obtain LPR status. Those immigrating as professional and skilled workers face wait times of many years, but those who meet the criteria of the extraordinary ability or advanced degrees preference categories have a much shorter wait.

STEM visas have gained interest in the 112th Congress, and various bills with STEM visa provisions (H.R. 399, H.R. 2161, H.R. 3146, H.R. 5893, H.R. 6412, S. 1965, S. 1986, S. 3185, S. 3192, and S. 3217) have been introduced. The House Committee on the Judiciary held two hearings on STEM and other high-skilled immigration in 2011. These issues also arose during a 2011 Senate Committee on the Judiciary hearing on the economic rationale for immigration reform. On September 20, 2012, the STEM Jobs Act of 2012 (H.R. 6429) failed to receive the necessary two-thirds vote to pass under suspension of the rules. Most recently, the House passed a revised version of the STEM Jobs Act of 2012 (H.R. 6429) on November 30, 2012.

Congress has renewed its interest in facilitating the immigration of foreign professional workers in science, technology, engineering, or mathematics (STEM) fields. The STEM workforce is seen by many as a catalyst of U.S. global economic competitiveness and is likewise considered a key element of the legislative options aimed at stimulating economic growth.[1] "STEM visa" is a shorthand for an expedited immigration avenue that enables foreign nationals with graduate degrees in STEM fields to adjust their immigration status to legal permanent residence (LPR) without waiting in the queue of numerically-limited LPR visas.[2] The fundamental policy question is: should the United States create additional pathways for STEM graduates to remain in the United States permanently? Or, are current avenues adequate?

The answer to the question lies at the nexus of education policy, labor force needs, and immigration priorities. More precisely, the key elements are: the source countries of international STEM students; the hiring choices of U.S. employers; and, the statutory limits and priorities of U.S. immigration law. This report opens by presenting a statistical portrait of foreign nationals studying STEM fields in U.S. institutions. An analysis of the current avenues foreign nationals with STEM degrees use to work in the United States temporarily and permanently follows. Discussions of the legislative history, current legislation, and major issues of debate conclude the report.

BACKGROUND

The Immigration and Nationality Act (INA) currently allocates 140,000 visas annually for economic immigrants, of which over 120,000 are allocated to three employment-based preference categories. These employment-based LPR categories are (1) persons of extraordinary ability in the arts, sciences, education, business, or athletics; outstanding professors and researchers; and certain multinational executives and managers; (2) members of the professions holding advanced degrees or persons of exceptional ability; and (3) skilled workers with at least two years training, professionals with baccalaureate degrees, and unskilled workers in occupations in which U.S. workers are in short supply.[3]

Although the United States remains the leading host country for international students in STEM fields, the global competition for talent, most notably with the European Union and Asian nations, has intensified in recent years. Concerns that these educational and career opportunities foster a "brain drain" of professionals from developing nations are infrequently expressed in the current debate. Instead, some researchers warn of a "reverse brain drain" because the United States does not allocate what they consider to be a sufficient number of visas for high-skilled immigrants to fuel the economy. According to the U.S. Department of Commerce, "growth in STEM jobs was three times as fast as growth in non-STEM jobs" over the past 10 years.[4]

Other researchers maintain that a record number of STEM graduates—both U.S. residents and foreign nationals—are entering the U.S. labor market. They express concern that foreign nationals would displace U.S. residents in the STEM fields if additional visas were allocated. Some analysts observe that the only high-skilled occupations that experienced negative wage growth in recent years were technology-related occupations (e.g., computer programmers and engineers)—occupations in which highly-educated foreign nationals cluster. Almost two-thirds of the 9.3 million people in the U.S. labor market who had STEM degrees in 2010 were employed in non-STEM occupations.[5]

Caveat

There is no generally accepted definition of what specific academic disciplines "STEM" encompasses. The National Science Foundation (NSF) studies the fields broadly and includes biological, agricultural, and environmental life sciences; computer and information sciences; mathematics and statistics; the physical sciences; psychology; the social sciences; engineering; and health fields.[6] For the purposes of tracking foreign students who study in the United States, the Department of Homeland Security's (DHS's) Immigration and Customs Enforcement (ICE) defines STEM more narrowly and does not include economics, sociology, and political science; however, ICE recently expanded its list, adding fields such as animal science, educational statistics, environmental science, nutritional science, and various specializations within psychology.[7] The National Center for Education Statistics often uses the ICE definition.[8]

TEMPORARY FOREIGN STUDENTS

International students pursuing STEM degrees in the United States are part of a long tradition of foreign students studying at U.S. institutions of higher education. Since the Immigration Act of 1924, the United States has expressly permitted foreign students to study in U.S. institutions. While their presence is generally viewed as a positive one, Congress has strengthened the reporting requirements of educational institutions with foreign students.[9] Foreign students are among the various categories of foreign nationals who are known as nonimmigrants.[10]

The most common visa for foreign students is the F-1 visa. It is tailored for international students pursuing a full-time academic education.[11] Spouses and children may accompany the F-1 visa holders on F-2 visas, but are not permitted to work. Students on F visas are generally barred from off-campus employment; however, they are permitted to work in practical training that relates to their degree program, such as paid research and teaching assistantships.[12] The Department of Homeland Security (DHS) Student and Exchange Visitor Information System (SEVIS) database reports 512,884 F-1 students were in active status (i.e., enrolled full-time or otherwise engaged in approved activities) in FY2010.[13]

Foreign Nationals Earning STEM Degrees

The number of full-time graduate students in science, engineering, and health fields who were foreign students grew from 91,150 in 1990 to 148,923 in 2009. Based on the more broadly defined NSF data, **Figure 1** shows that most of the increase occurred after 1999. Despite the rise in foreign student enrollment, the percentage of STEM graduate students with temporary visas in 2009 (32.7%) was comparable to 1990 (31.1%). Graduate enrollments in engineering fields have exhibited the most growth of STEM fields in recent years.[14] Data from the National Center for Education Statistics' Integrated Postsecondary Education Data System (IPEDS) indicate that 10,000 foreign students earned doctoral degrees in STEM fields in 2009 and almost 30,000 foreign students earned masters degrees in STEM fields in 2009.[15]

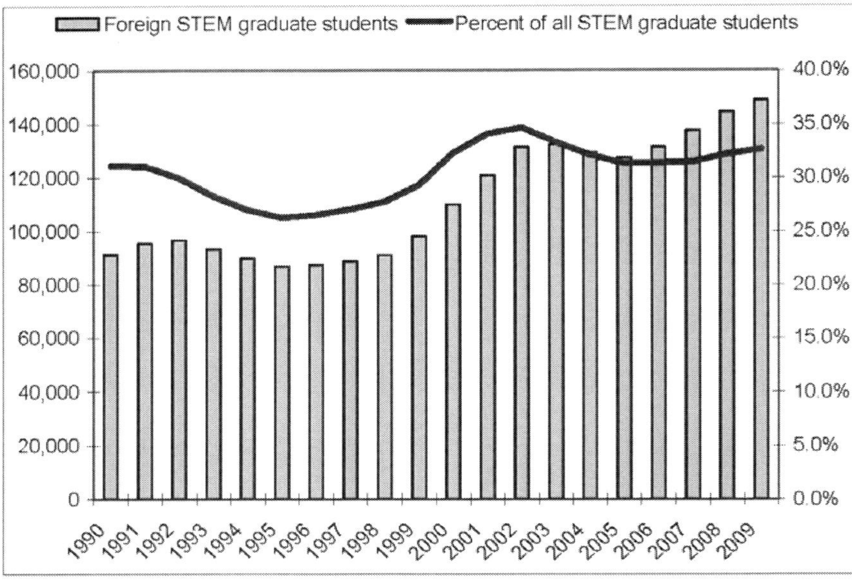

Source: National Science Foundation/National Center for Science and Engineering Statistics, NSF-NIH Survey of Graduate Students and Postdoctorates in Science and Engineering.

Figure 1. Full-Time Graduate Students with Temporary Visas in Science, Engineering, and Health Fields, 1990-2010.

A snapshot of degrees granted to foreign students by broad categories indicates that engineering is the leading field for both MA and Ph.D. recipients in 2010 (**Figure 2**). Computer science fields follow for MA degrees; while mathematics and physical sciences place second for Ph.D. degrees. Although based upon data from ICE, the STEM fields presented in **Figure 2** use the broader NSF definition of STEM, which includes economics, sociology, psychology and political science as well as agricultural and life sciences, computer sciences, mathematics and the physical sciences.

India was the top sending country for STEM graduates enrolled in masters degree programs and represented 56% of all STEM students seeking masters degrees in 2009 (**Table 1**). Indian graduate students studying engineering and computer science led in masters degrees. Students from China placed second at 15% and were more evenly distributed across STEM fields, with the exception of engineering. The remaining top eight sending countries accounted for a total of 13% of all foreign students seeking masters degrees in STEM fields in 2009, less than that of China.

In terms of doctoral degrees (**Table 2**), China sent the most STEM students in 2009, making up 35%. Chinese Ph.D. students were almost half of all foreign nationals in mathematics and physical sciences and roughly a third of all foreign nationals in the other STEM fields (except psychology and the social sciences). Doctoral degree students from India represented 16% of STEM doctorates in 2009.[16]

Table 1. Foreign Nationals Enrolled in STEM Fields, Master's Degrees 2009 Top 10 sending countries by broad categories

	Agricultural Biological Sciences	**Computer Science**	**Economics**	**Engineering**	**Mathematics Physical Sciences**	**Psychology Social Sciences**	**Percent of Total**
Total Master's	6,050	22,880	2,480	40,300	5,810	7,600	
India	2,750	16,270	140	26,290	1,210	500	56%
China	1,030	2,240	1,040	5,100	2,290	1,220	15%
Taiwan	240	520	100	1,240	200	340	3%
South Korea	140	440	140	910	280	670	3%
Nepal	160	460	50	430	110	100	2%
Saudi Arabia	70	320	40	380	90	110	1%
Canada	150	70	30	210	70	470	1%
Japan	90	70	40	130	70	520	1%
Turkey	30	130	80	380	60	190	1%
Thailand	40	250	40	330	40	90	1%

Source: Department of Homeland Security, U.S. Immigration and Customs Enforcement, Student and Exchange Visitor Information System database, special tabulations for the National Science Foundation, 2010.

Note: Although based upon data from ICE, these STEM fields use the broader NSF definition of STEM.

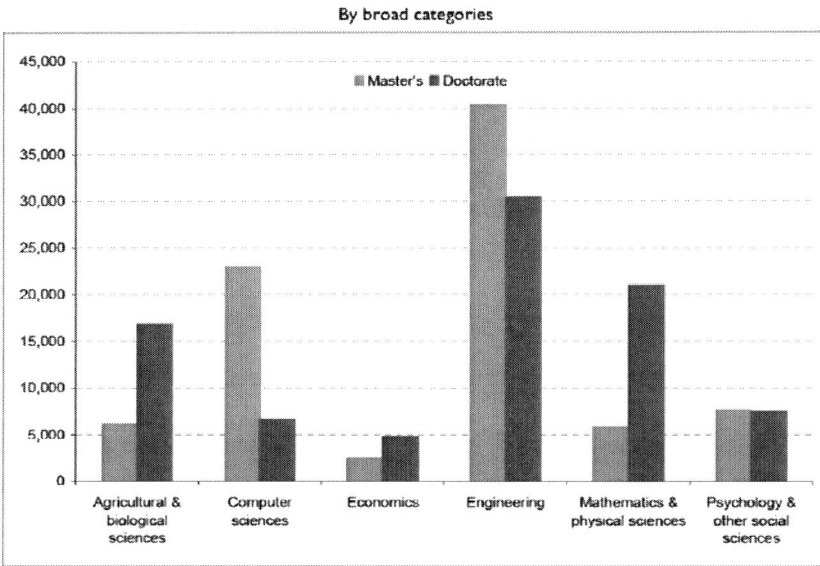

Source: Department of Homeland Security, U.S. Immigration and Customs Enforcement, Student and Exchange Visitor Information System database, special tabulations for the National Science Foundation, 2010.

Note: Although based upon data from ICE, these STEM fields use the broader NSF definition of STEM.

Figure 2. Foreign Nationals Enrolled in Advanced Degrees in STEM Fields, 2009.

Table 2. Foreign Nationals Enrolled in STEM Fields, Doctorate Degrees 2009 Top 10 sending countries by broad categories

	Agricultural Biological Sciences	Computer Science	Economics	Engineering	Mathematics Physical Sciences	Psychology Social Sciences	Percent of Total
Total Doctorates	16,830	6,650	4,800	30,380	20,900	7,490	
China	5,550	2,410	1,300	10,570	8,680	980	35%
India	3,600	1,250	370	5,690	2,800	530	16%
South Korea	1,080	520	530	3,100	1,240	1,070	9%
Taiwan	930	220	150	1,480	700	400	4%
Turkey	250	210	280	980	440	460	3%
Iran	90	210	60	1,830	270	50	3%

	Agricultural Biological Sciences	Computer Science	Economics	Engineering	Mathematics Physical Sciences	Psychology Social Sciences	Percent of Total
Canada	460	90	80	430	380	690	2%
Japan	240	30	160	160	210	340	1%
Germany	210	70	100	150	250	210	1%
Nepal	200	40	60	160	410	50	1%

Source: Department of Homeland Security, U.S. Immigration and Customs Enforcement, Student and Exchange Visitor Information System database, special tabulations for the National Science Foundation, 2010.

Note: Although based upon data from ICE, these STEM fields use the broader NSF definition of STEM.

Temporary Foreign Workers

The Immigration and Nationality Act of 1952 authorized visas for foreign nationals who would perform needed services because of their high educational attainment, technical training, specialized experience, or exceptional ability. Today, there are several temporary visa categories that enable employment-based temporary admissions for highly skilled foreign workers. They perform work that ranges from professional and skilled labor to those who work in jobs requiring extraordinary ability in the sciences, arts, education, business, or athletics.

Optional Practical Training (OPT)

After completing their undergraduate or graduate studies, F-1 foreign students are permitted to participate in employment known as Optional Practical Training (OPT), which is temporary employment that is directly related to an F-1 student's major area of study. Generally, an F-1 foreign student may work up to 12 months in OPT status. In 2008, DHS expanded the OPT work period to 29 months for F-1 students in STEM fields. To qualify for the 17-month extension, F-1 students must have received STEM degrees included on the STEM Designated Degree Program List, be employed by employers enrolled in E-Verify,[17] and have received an initial grant of post-completion OPT related to such a degree (i.e., already approved for 12 months in OPT).[18]

According to U.S. Citizenship and Immigration Service (USCIS), the number of F-1 visa holders who are engaged in OPT was 92,465 in FY2010. They comprise 18% of all active F-1 foreign students in FY2010. STEM graduates performing OPT during their first year after graduation are included among the 82,206 F-1 students shown in **Figure 3** as "Post-Completion." In addition, there were 10,022 F-1 students who obtained the 17-month extension for OPT in STEM fields.

Temporary Professional Specialty Worker: H-1B Visas

Many F-1 visa holders (especially those who are engaged in OPT) often change their immigration status to become professional specialty workers (H-1B). Other H-1B visa holders, however, are adjusting from another immigration status or coming directly from abroad. The law sets numerical restrictions on annual admissions of the H-1B (65,000). Most H-1B workers, however, enter on visas that are exempt from the ceiling because they are returning workers or they work for universities and nonprofit research facilities that are exempt from the cap. The H-1B visas are closely associated with STEM fields, but are not limited to them.

Prospective H-1B nonimmigrants must demonstrate to USCIS that they have the requisite education and work experience for the posted positions. After the Department of Labor (DOL) approves the labor attestation, USCIS processes the petition for the H-1B nonimmigrant (assuming other immigration requirements are satisfied) for periods up to three years. An alien can stay a maximum of six years on an H-1B visa.

The H-1B labor attestation, a three-page application form, is the first step for an employer wishing to bring in an H-1B professional foreign worker. The attestation is a statement of intent rather than a documentation of actions taken.[19] In the labor attestation for an H-1B worker, the employer must attest that the firm will pay the nonimmigrant the greater of the actual wages paid to other employees in the same job or the prevailing wages for that occupation; that the firm will provide working conditions for the nonimmigrant that do not cause the working conditions of the other employees to be adversely affected; and that there is no applicable strike or lockout. The firm must provide a copy of the labor attestation to representatives of the bargaining unit or—if there is no bargaining representative—must post the labor attestation in conspicuous locations at the work site.[20]

In FY2011, there were 218,500 H-1B professional specialty worker petitions approved, down from a high of 288,000 in FY2004. Although the law sets a numerical limit of 65,000 H-1B workers each year, only initial grants are counted under the cap. As noted above, there are several categorical exceptions to the cap, most notably all H-1B nonimmigrants who work for universities and nonprofit research facilities. A provision in P.L. 108-447 also exempts up to 20,000 aliens holding a master's or higher degree from the numerical limit on H-1B visas. As **Figure 4** displays, over the past decade more H-1B workers were approved outside of the numerical cap of 65,000 than under the cap. Not all H-1B workers with approved petitions actually use the visa.

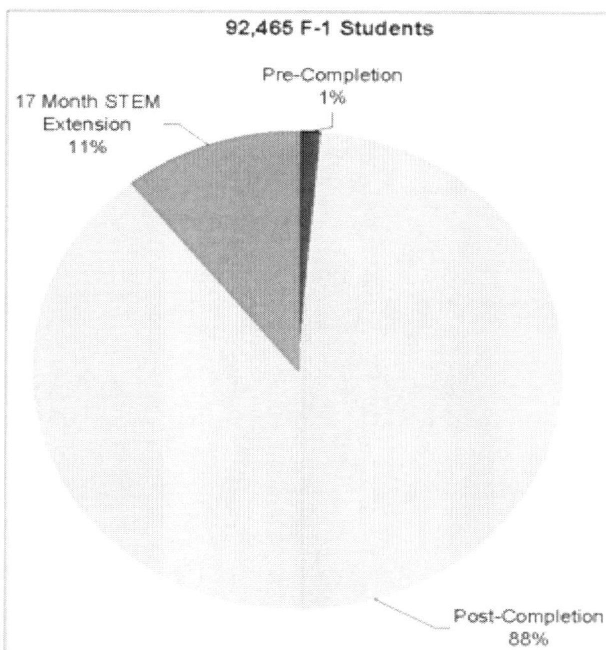

Source: CRS presentation of data from U.S. Citizenship and Immigration Services.

Notes: Post-Completion category includes those STEM graduates in the first year of OPT.

Figure 3. F-1 Foreign Students Performing Optional Practical Training in FY2010.

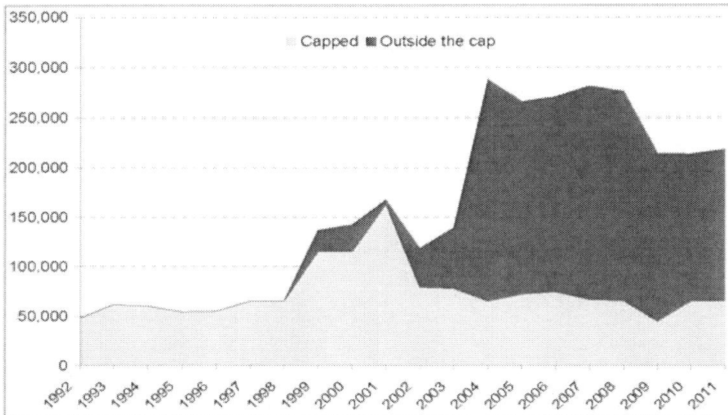

Source: CRS presentation of data from the DHS Office of Immigration Statistics (OIS) and its predecessor in the Immigration and Naturalization Service.

Notes: Congress increased the H-1B cap to 115,000 for FY1999-FY2000 and to 195,000 for FY2001-FY2003.

Figure 4. Total H-1B Petitions Approved, FY1992-FY2011.

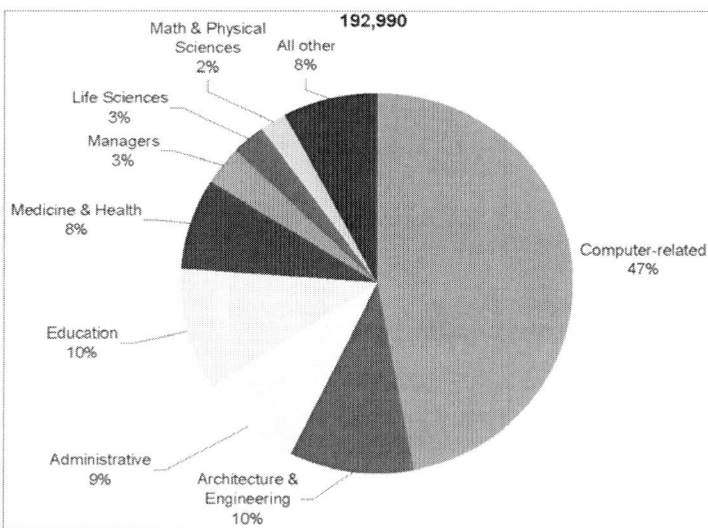

Source: CRS presentation of data from U.S. Citizenship and Immigration Services.

Note: H-1B beneficiaries include workers renewing their visas as well as newly arriving workers.

Figure 5. Occupations of H-1B Worker Beneficiaries in FY2010.

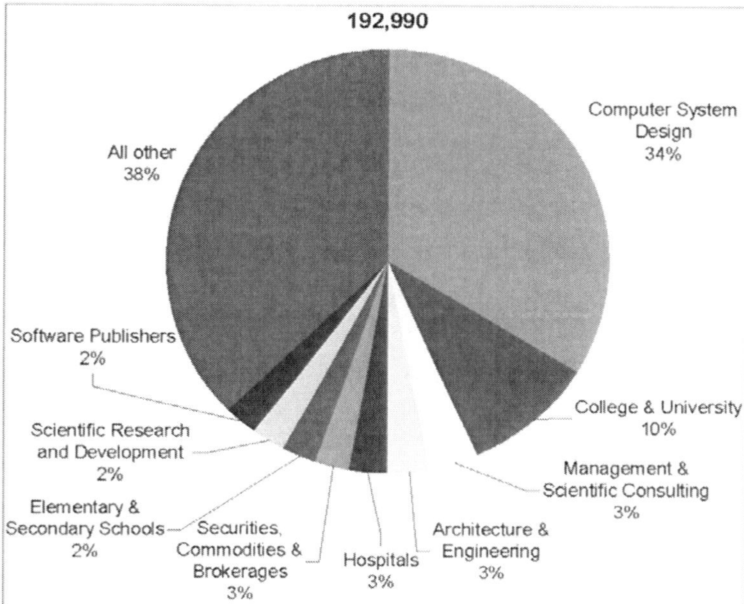

Source: CRS presentation of data from U.S. Citizenship and Immigration Services.

Note: H-1B beneficiaries include workers renewing their visas as well newly arriving workers.

Figure 6. Top Trade/Industry Sectors Hiring H-1B Worker Beneficiaries, 2010.

Over the years, a noteworthy portion of H-1B beneficiaries have worked in STEM occupations. In FY2010, the most recent year for which detailed data on H-1B beneficiaries (i.e., workers renewing their visas as well newly arriving workers) are available, almost 91,000 H-1B workers were employed in computer-related occupations, and they made up 47% of all H-1B beneficiaries that year, as **Figure 5** indicates. Architectural and engineering occupations as well as occupations in education were tied at a distant second with 10% each. Administrative occupations followed with 9%, and health and medicine occupations were 8% of the 192,990 H-1B beneficiaries. The total number of H-1B beneficiaries reported for FY2010 (192,990) and shown in **Figure 5** was less than the number of approved H-1B petitions approved that year, as depicted in **Figure 4**.[21]

In terms of the broad trade group categories in which the H-1B beneficiaries were employed in FY2010, the leading trade group was computer systems design, which employed about a third (34%) of the H-1B workers

(**Figure 6**). Colleges and universities employed 10% of the H-1B beneficiaries. Similar numbers of H-1B beneficiaries were employed in the following trade/industry sectors: management and scientific consulting; architecture and engineering; hospitals; and, securities, commodities, and brokerages. Each of these groups have hired about 3% of the H-1B beneficiaries. **Figure 6** shows that a noteworthy portion of the H-1B beneficiaries (38%) were not employed in one of the top trade/industry sectors, suggesting they are employed across the various trade/industry sectors.[22]

Other Professional Specialty Workers: TN and E-3 Visas

There are two nonimmigrant visa categories quite similar to H-1B visas that are designated for temporary professional workers from specific countries. These visas are based upon specific trade agreements foreign nations have signed with the United States. Canadian and Mexican temporary professional workers may enter according to terms set by the North American Free Trade Agreement (NAFTA) on TN visas. The E-3 treaty professional visa is a temporary work visa limited to citizens of Australia.[23] Occupationally, they mirror the H-1B visa in that the foreign worker on an E-3 visa or a TN visa must be employed in a specialty occupation. They do not, however, require a labor attestation.

There are several other employment-based nonimmigrant visas for which foreign students with STEM degrees may be eligible. These nonimmigrants include cultural exchange workers on J visas; multinational executive and specialist employees on L visas; international investors on E visas; and persons with outstanding and extraordinary ability on O visas. For further information, see the Appendix, "Other High-Skilled Temporary Employment Categories."

Pathways to Legal Permanent Residence

The H-1B visa and OPT often provide the link for foreign students to become employment-based LPRs. Many anecdotal accounts tell of foreign students who are hired by U.S. firms as they are completing their programs. The employers obtain H-1B visas for the recent graduates or hire them as OPT to extend their F-1 visas. According to DHS: "This extension of the OPT period for STEM degree holders gives U.S. employers two chances to recruit these highly desirable graduates through the H-1B process, as the extension is long enough to allow for H-1B petitions to be filed in two successive fiscal years."[24] If the temporary foreign workers meet expectations, the employers may also petition for them to become LPRs through one of the employment-based immigration categories.[25]

Over 90% of 140,000 employment-based LPRs annually are adjusting to LPR status within the United States rather than newly arriving from abroad. Presumably, many of these foreign nationals had originally entered the United States as foreign students or temporary workers. Most foreign nationals seeking to qualify for a nonimmigrant visa, however, must demonstrate that they are not coming to reside permanently. Temporary workers who are H-1B visa holders are permitted to petition for a LPR visa at the same time that they file for an H-1B visa, a policy exception known as dual-intent.[26]

Permanent Employment-Based Admissions

Admissions and adjustments for LPR status are subject to a complex set of numerical limits and preference categories that give priority for admission on the basis of family relationships, needed skills, and geographic diversity. The INA establishes an annual worldwide level of 140,000 employment-based preference immigrants, and that ceiling includes the accompanying spouse and children in addition to the principal (i.e., qualifying) alien.[27] The employment-based preference categories are as follows:

- *first preference*: 40,040 for priority workers who are persons of extraordinary ability in the arts, sciences, education, business, or athletics; outstanding professors and researchers; and certain multinational executives and managers;
- *second preference*: 40,040 for members of the professions holding advanced degrees or persons of exceptional ability;
- *third preference*: 40,040 skilled workers with at least two years training, professionals with baccalaureate degrees, and unskilled workers in occupations in which U.S. workers are in short supply;
- *fourth preference*: 10,000 for special immigrants who largely consist of religious workers, certain former employees of the U.S.

government, and undocumented juveniles who become wards of the court; and

- *fifth preference*: 10,000 for investors who invest at least $1 million (or less money in rural areas or areas of high unemployment) to create at least 10 new jobs.[28]

The employment-based visas made up 14.2% of the total of 1 million LPRs in FY2010.[29]

Foreign nationals with STEM degrees may qualify under several of the preference categories, depending on their talent, educational attainment, expertise, and experience. Employers who seek to hire prospective employment-based immigrants through the second and third preference categories also must petition DOL on behalf of the alien. The prospective immigrant must demonstrate that he or she meets the qualifications for the particular job as well as the preference category. If DOL determines that a labor shortage exists in the occupation for which the petition is filed, labor certification will be issued. If there is not a labor shortage in the given occupation, the employer must submit evidence of extensive recruitment efforts in order to obtain certification.[30]

From FY2000 through FY2009, 145,390 foreign nationals qualified as first preference principals (i.e., priority workers who are persons of extraordinary ability, outstanding professors and researchers or certain multinational executives and managers). Of these, 21% reported they were engaged in STEM occupations, using the broader NSF definition. As **Figure 7** depicts, almost half (49%) of the first preference principals reported executive, administrative or managerial professions.[31]

As **Figure 7** illustrates, 61% of the second preference principals reported they were engaged in STEM occupations during FY2000-FY2009. The broader NSF definition of STEM fields was used for this analysis. In other words, STEM occupations dominated among the 184,382 foreign nationals who were admitted over this decade as members of the professions holding advanced degrees or persons of exceptional ability.

STEM occupations also made up a plurality of the third preference principals (i.e., skilled workers with at least two years training, professionals with baccalaureate degrees, or unskilled workers in occupations in which U.S. workers are in short supply). From FY2000 through FY2009, 45% of the 346,870 foreign nationals admitted as third preference principals reported they worked in STEM occupations.

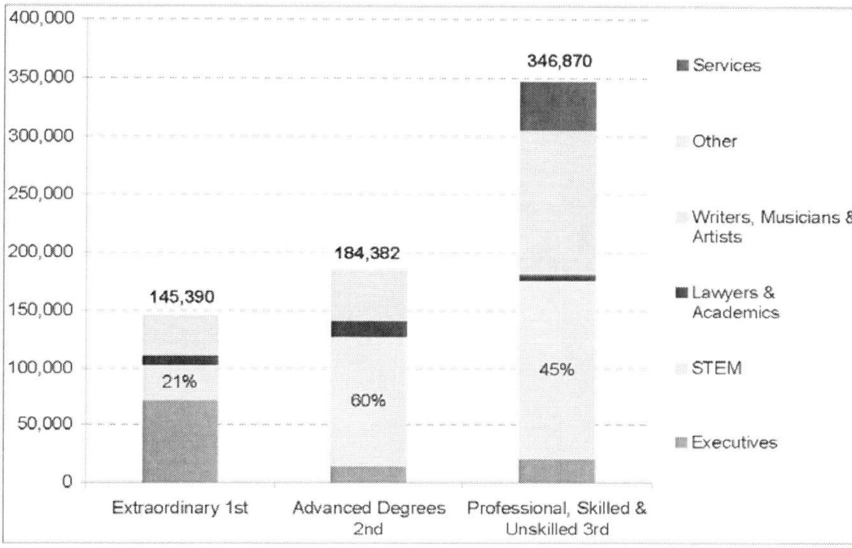

Source: CRS analysis of data from the DHS Office of Immigration Statistics and the former Immigration and Naturalization Service.

Notes: STEM includes biological, agricultural, and environmental life sciences; computer and information sciences; mathematics and statistics; the physical sciences; psychology; the social sciences; engineering; and health fields.

Figure 7. Occupations of 1st through 3rd Preference Employment-Based Principals, FY2000-FY2009.

In total, foreign nationals reporting STEM occupations made up 44% of all of the 676,642 LPRs who were employment-based principals in the first, second, and third preference categories during the decade of FY2000-FY2009. Of all of the LPRs reporting STEM occupations (297,668) over this decade, 52% entered as third preference principals.

Employment-Based I-485 Applications Pending

The volume of employment-based immigration cases pending is an important factor in the STEM debate. The USCIS maintains a system of approved employment-based I-485 applications (i.e., the Application to Register Permanent Residence or Adjust Status) that are pending, which provides another source of data on the number of approved employment-based

LPRs. Known as the I-485 Inventory, these data are available by preference category and by top countries. These I- 485 data include the employment-based applicants who plan to adjust their immigration status within the United States. The prospective employment-based LPRs who would be new arrivals from abroad are not included in the I-485 inventory, because they would not need to file I-485 petitions.

As **Figure 8** shows, STEM graduates seeking LPR status are likely to wait in line to obtain LPR status, but those who meet the criteria of the extraordinary ability or advanced degrees preference categories have a much shorter wait. That there are no third preference I-485 applications filed after August 2007 in the I-485 Inventory suggests that USCIS has not been approving many since the 2007 visa retrogression pushed back the visa priority dates. These data likely understate the number of cases pending.[32]

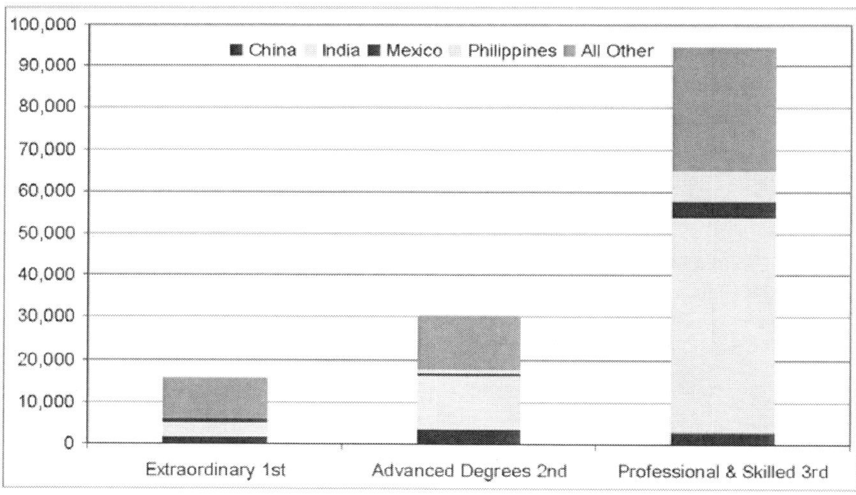

Source: USCIS I-485 Inventory of pending cases, as of January 5, 2012.

Note: There were 140,245 cases; however, no "Professional and Skilled" (3rd) preference applications filed after August 2007 are included in the I-485 Inventory thus far.

Figure 8. Employment-Based I-485 Applications Pending January 2012 by Preference and Top Countries.

LEGISLATIVE HISTORY OF STEM VISAS

Legislative proposals for STEM visas emerged during the debate over Comprehensive Immigration Reform in the mid-2000s. Various options aimed at retaining STEM graduates of American universities were folded into the legislation that was moving at the time. The approach was two-fold: (1) foster the recruitment and retention of STEM foreign students; and (2) enable foreign nationals with graduate degrees in STEM fields to adjust to LPR status without waiting in the queue of numerically limited visas. Although the ideas for STEM visas continue to be discussed, legislative action on STEM visas stalled along with the broader Comprehensive Immigration Reform efforts in 2007.

During the 109th Congress, the Senate-passed Comprehensive Immigration Reform legislation (S. 2611) was the first piece of legislation with STEM visa provisions that received action. The legislation would have exempted from numerical limits foreign nationals who had worked in the United States for three years and who had earned an advanced degree in a STEM field. More specifically, §507 of S. 2611 as passed by the Senate would have permitted foreign students on a proposed STEM student visa to adjust to LPR status. Also, §508 would have allowed an unlimited number of foreign nationals who had earned an advanced degree in a STEM field and had been working in a related field in the United States during the three years preceding to become LPRs. The 109th House of Representatives did not pass comparable legislation.

In the 110th Congress, S.Amdt. 1150 to S. 1348, the Comprehensive Immigration Reform legislation that failed a cloture vote in the Senate, as well as H.R. 1645 (Security Through Regularized Immigration and a Vibrant Economy Act of 2007, or the STRIVE Act), similarly would have created STEM visas. The legislative proposals in the 110th Congress would have established a STEM foreign student visa category and would have permitted an unlimited number of foreign nationals who had earned STEM degrees and had been working in a related field in the United States during the three years preceding to become LPRs. Foreign students seeking STEM visas would not have needed to demonstrate an intent of departing the United States upon completion of their studies. Students in this category would have also been permitted to pursue Optional Practical Training for periods of up to 24 months after completing their degree. Employers would have been required to pay foreign students the higher of the average or prevailing wage in the field of employment.

SELECTED LEGISLATION IN THE 112TH CONGRESS

There has been renewed interest in establishing STEM visas in the 112th Congress, and several bills have been introduced in both chambers. The House Committee on the Judiciary held two hearings on these issues in 2011.[33] These issues also arose during the Senate Committee on the Judiciary hearing on the economic rationale for immigration reform.[34] A variety of bills creating STEM visas have been introduced, including H.R. 399, H.R. 2161, H.R. 3146, H.R. 5893, H.R. 6412, H.R. 6429, S. 1965, S. 1986, S. 3185, S. 3192, and S. 3217.[35]

The Immigration Driving Entrepreneurship in America Act of 2011, or the IDEA Act of 2011 (H.R. 2161); the American Innovation and Education Act of 2011 (H.R. 3146); and the Science, Technology, Engineering, and Mathematics Visa Act of 2011, or STEM Visa Act of 2011 (S. 1986), while not identical, do include immigration provisions that are comparable in terms of STEM visas. The bills would amend the INA to establish a priority worker immigrant visa for an alien who has a qualifying: (1) master's or higher degree in a field of science, technology, engineering, or mathematics (STEM degree) from a U.S. institution of higher education; and (2) employment offer from a U.S. employer. S. 1986, H.R. 2161, and H.R. 3146 would also eliminate the foreign residency requirement for certain foreign students, enabling them to have dual intent.[36] The bills further would allow extensions of stay based upon a lengthy labor certification or priority adjudication for F-1 students, H-1B workers, and L intracompany transfers.

S. 1986, H.R. 2161 and H.R. 3146 would also amend the INA to impose a fee of $2,000 per petition on an employer who files an application to hire a foreign worker. Certain not-for-profit and university-based employers would be exempt from paying the fee. Funds from the fees collected would be deposited in the STEM Education and Training Account, which the bills would establish in the Treasury. S. 1986, H.R. 2161 and H.R. 3146 furthermore would require that 60% of the funds be used for a STEM scholarship program for low-income U.S. students enrolled in degree programs in science, technology, engineering, or mathematics.[37]

The Stopping Trained in America Ph.D.s From Leaving the Economy Act of 2011 or the STAPLE Act (H.R. 399) would amend the INA to exempt (1) from direct numerical limitations aliens who have earned a Ph.D. degree in a STEM field from a U.S. institution of higher education and who have an offer of employment from a U.S. employer in a field related to such degree; and (2) from H-1B visa numerical limitations aliens who have earned a Ph.D. degree in a STEM field from a U.S. institution, so long as that degree is required for the job the petitioning employer is filling.

The Startup Act of 2011[38] (S. 1965) and the Startup 2.0 Act of 2012 (H.R. 5893/S. 3217) would create a new conditional resident visa for up to 50,000 foreign students who have a master's or higher degree in a STEM field from a U.S. institution of higher education. The recipients of these new STEM visas would be granted conditional residence status contingent upon them remaining actively engaged in a STEM field for five consecutive years, after which they would be able to adjust to legal permanent resident status.

Rather than adding LPR visas for STEM graduates, the Securing the Talent America Requires for the 21st Century Act of 2012, or STAR Act (S. 3185), would eliminate the Diversity Visa Lottery and re-allocate those 55,000 Diversity Visas to a new category it would create for foreign STEM graduates who have job offers in related fields.[39] In addition, S. 3185 would amend the labor certification requirements in the INA to create a special provision in the law for employers of foreign STEM master's degree recipients, which would require the employers to engage in a competitive recruitment and selection process and determine that the alien is more qualified than any U.S. worker who applied for the job. S. 3185 would further amend the INA so that job openings in which employers are petitioning to hire foreign STEM graduates with doctoral degrees would automatically be deemed as shortage occupations that do not require certification from the Department of Labor.

The Attracting the Best and the Brightest Act of 2012 (H.R. 6412) would add 55,000 visas for foreign STEM graduates who have job offers in related fields, but it would not eliminate the Diversity Visa Lottery. Any visas unused by the foreign STEM graduates would roll over to the other employment-based visa categories. It would define STEM as a field included in the Department of Education's Classification of Instructional Programs taxonomy within the summary groups of computer and information sciences and support services, engineering, mathematics and statistics, and physical sciences. H.R. 6412 would require foreign STEM graduates to have an advanced degree from an accredited public or nonprofit university classified by the National Science Foundation as either a research institution or as otherwise excelling in STEM instruction. In provisions similar to the STEM Jobs Act of 2012 (discussed below), H.R. 6412 would amend the labor certification provisions in the INA to require employers of foreign STEM graduates to submit a job order for the position with the state workforce agency in that state, which in turn is required to post the position on its official website. However, H.R. 6412 would also require that employers offer wages to the foreign STEM graduates that meet or exceed the actual wages paid to U.S. workers with similar levels of experience.

The STEM Jobs Act of 2012 (H.R. 6429) would eliminate the Diversity Visa Lottery and re- allocate those 55,000 Diversity Visas to two new categories it would create for foreign STEM graduates who have job offers in related fields.[40] Though the STEM Jobs Act of 2012 would allocate all of the 55,000 visas to the first new category for foreign nationals who have earned a Ph.D. degree in a STEM field from a U.S. institution, any remaining visas would roll down to the second new category for foreign nationals who have earned a M.A. degree in a STEM field. Unlike H.R. 6412, however, any visas unused by the foreign STEM graduates would not roll over to the other employment-based visa categories under the STEM Jobs Act of 2012.

Similar to H.R. 6412, the STEM Jobs Act of 2012 would define STEM as a field included in the Department of Education's Classification of Instructional Programs taxonomy within the summary groups of computer and information sciences and support services, engineering, mathematics and statistics, and physical sciences. The foreign STEM graduates would have to receive degrees from a doctorate-granting university that the Carnegie Foundation for the Advancement of Teaching rates as having a very high or high level of research activity or that the National Science Foundation has classified as having research activity equivalent to those institutions classified by the Carnegie Foundation. It would amend the labor certification provisions in the INA to require employers of foreign STEM graduates to submit a job order for the position with the state workforce agency in that state, which in turn is required to post the position on its official website.

On September 20, 2012, the STEM Jobs Act of 2012 (H.R. 6429) failed to receive the necessary two-thirds vote to pass under suspension of the rules. The legislation garnered 257 yeas and 158 nays, with bipartisan support as well as bipartisan opposition.[41]

Most recently, the House passed revised version of the STEM Jobs Act of 2012 (H.R. 6429) that offers several key differences from the legislation that the House considered on September 20, 2012. The House-passed version of H.R. 6429

- would permit unused STEM visas made available in fiscal years 2013 through 2016 to be utilized in future years, under terms set by the bill;
- would eliminate a provision requiring STEM visa recipients to promise to work for at least five years in the United States or for at least five years for the petitioner in a STEM field; and,

- would eliminate a provision prohibiting universities recognized for prospective STEM visa recipients from offering incentive payments to persons based on securing foreign students.

Perhaps the most significant addition to H.R. 6429 is the expansion of the nonimmigrant visa for family members with approved LPR petitions pending, commonly known as the "V" visa.[42] The bill would allow the spouses and minor children of LPRs to live in the United States on "V" visas (without employment authorization) while they wait for their LPR visas to become available. Immediate relatives of LPRs would become eligible for the V visas after one year on the visa priority date waitlist.[43]

The House passed H.R. 6429 on November 30, 2012 by a vote of 245 Yeas and 139 Nays.[44]

Policy Discussion

The fundamental policy question for Congress is should the United States create additional pathways for STEM graduates to remain in the United States permanently? Some policymakers consider establishment of STEM visas to be a natural and positive chain of events, arguing that it would be inadvisable to educate talented young people only to make them leave to work for foreign competitors. Others maintain employment-based LPR admissions are highly skewed toward high-skilled immigrants, which already provide for generous flows of foreign nationals with STEM degrees. A corollary factor is that pulling prospective LPRs with graduate STEM degrees out of the numerically limited employment-based categories would free up visas for the other prospective LPRs waiting in the employment-based queue. The policy discussion opens with two competing perspectives on the immigration prospects of international students with STEM degrees. Four themes of debate conclude the report: assessments of U.S. labor market needs, the competing definitions of STEM fields, the role of colleges and universities as gatekeepers, and the context of broader immigration reforms.

Two Perspectives on STEM Immigration Prospects

At the crux of the legislative debate is whether expedited immigration pathways for STEM foreign graduate students would be in the national interest, and two widely cited studies of the issue are illustrative of this debate. The two studies featured below yield divergent findings, in large part because they use different research designs and study different STEM populations. This section does not purport to say that one research design is superior to another or that one target population is more meritorious than the others. Rather, it offers two examples with distinct approaches to highlight the competing perspectives. One approach is from the vantage point of the immigration caseloads in the visa categories that STEM degree holders currently use to become LPRs; it explores the question of how long would foreign nationals from China and India who have STEM degrees wait to become LPRs of the United States. The other approach is a retrospective analysis of the foreign nationals who earn doctorates in STEM fields; it addresses the issue of retention rates of foreign STEM doctorates in the United States 10 years after graduation.

STEM Graduates Face Long Waits for LPR Status

A 2011 study by Stuart Anderson of the National Foundation for American Policy (NFAP) examined prospective LPRs with approved petitions to immigrate who were waiting in the queue for a numerically limited visa. Since the official data sources do not provide a complete total of cases in the pipeline, NFAP estimated some of the missing components.[45] The NFAP analysis of the data indicate that the majority of employer-sponsored LPRs at the end of 2010 were from India and China.

NFAP's Anderson estimated that a professional worker from India who applies for a professional and skilled (third preference) visa would wait approximately 70 years to obtain LPR status.[46]

Figure 9 presents the NFAP estimates for the upper end (i.e., maximum number of years) that prospective LPRs have been waiting and might continue to wait for an LPR visa according to when they filed their petitions.

The NFAP study of the pending caseload also estimated that a professional worker from China who applies for a third preference visa would wait approximately 24 years to obtain LPR status.[47] As **Figure 9** shows, the NFAP projected further that countries other than India and China would have wait times for a third preference of four to six years. Anderson concluded "America would lose much talent as U.S.-based businesses would need to hire or place such skilled individuals abroad, rather than invest in a green card process likely to last decades."[48]

The NFAP study indicates that foreign nationals holding advanced degrees (second preference) face far shorter wait times. Only India and China are mentioned as having a noteworthy number of second preference cases pending. Indians and Chinese seeking second preference LPR visas today would have waits of approximately six years, according to the NFAP study.[49]

NFAP's Anderson estimates that an exemption of 50,000 visas a year for foreign students with STEM graduate degrees would eliminate the backlog of cases for those holding advanced degrees (second preference) within two years. He further projects that it would ease the backlog of professional and skilled (third preference) workers, eliminating the backlog within 10 years.[50]

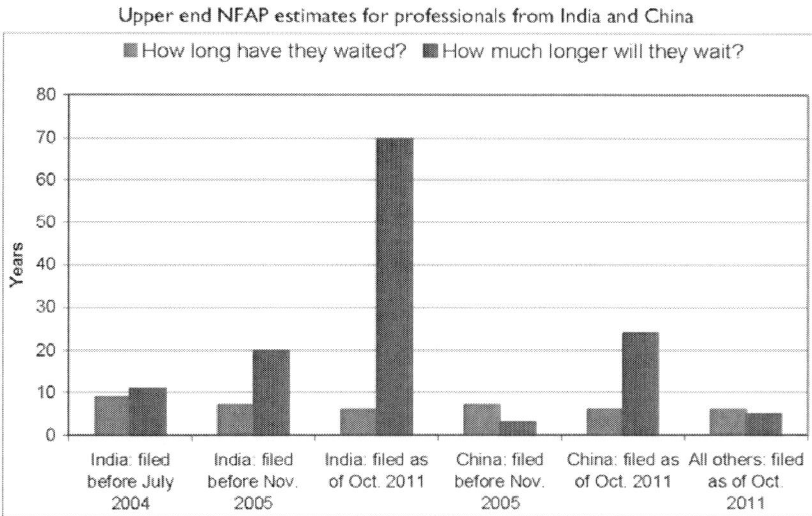

Source: *Waiting and More Waiting: America's Family and Employment-based Immigration System*, Table 2, National Foundation for American Policy (NFAP), October, 2011.

Notes: Bars depict the upper end (i.e., maximum number of years) of the NFAP estimates.

Figure 9. Projected Wait Times for Third Preference LPRs.

Stay Rates of STEM Graduates Remain High

Ongoing research that Michael G. Finn has conducted for the National Science Foundation (NSF) on foreign nationals receiving science and engineering doctorates 10 years after graduation offers a different perspective. According to his latest published analysis, the 2009 stay rate for all foreign doctorate recipients was 64% for those graduating five years earlier and 66% for those graduating 10 years earlier. **Figure 10** has been reproduced from his report.[51]

Finn found that the 2009 stay rate of doctorate recipients on temporary resident visas at the time of graduation behaved slightly differently for different cohorts. For those graduating five years earlier, the stay rate was down slightly from that recorded two years ago. However, for those graduating 10 years earlier, the stay rate in 2009 reached an all-time high. Finn concluded that stay rates for temporary foreign nationals receiving science and engineering doctorates overall have never been higher.[52]

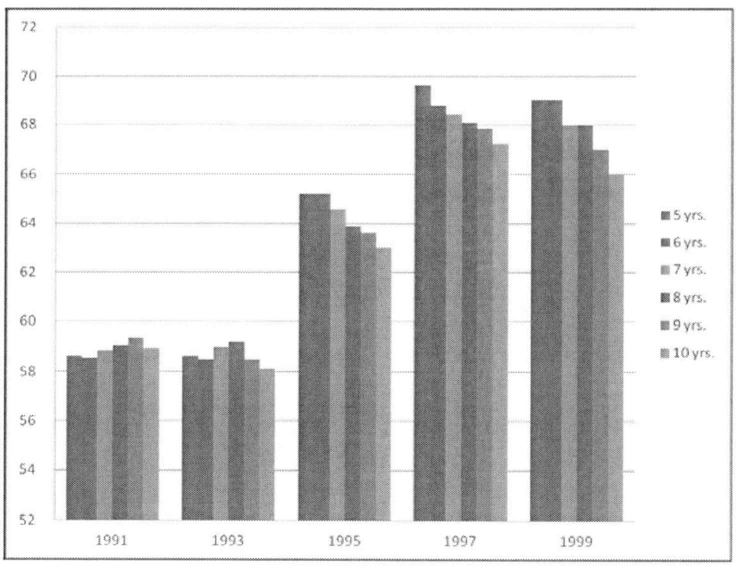

Source: reproduced from *Stay Rates of Foreign Doctorate Recipients from U.S. Universities, 2009*, by Michael G. Finn, January 2012.

Figure 10. Percentage of Foreign Nationals Receiving Science and Engineering Doctorates Who Were in the United States Five to 10 Years After Receipt of Doctorate, for Doctorates Awarded in 1991, 1993, 1995, 1997, and 1999.

Finn also studied prospective immigrants from India and China, and he reached this conclusion:

> China and India are countries of special interest because they account for a large and growing share of new doctorate recipients and are subject to some restrictions not faced by most other countries when seeking permanent resident status. When comparing the history of stay rates for these two countries with that of all other countries, however, there is no apparent evidence that visa restrictions have reduced stay rates for China and India.[53]

At first glance, it seems difficult to reconcile Finn's conclusion about prospective LPRs from China and India with Anderson's estimates of how long prospective LPRs from China and India might wait before visas would be available for them. From Anderson's perspective, the wait times are too long for prospective employment-based immigrants in the professional and skilled category. From Finn's perspective, those with science and engineering doctorates have sufficient time and opportunity to become LPRs through one of the extraordinary ability or advanced degrees categories.

U.S. Labor Market Needs

A broad consensus of business, academic, and policy leaders warn that the United States is on the verge of STEM workforce shortages, which will diminish U.S. global economic competitiveness. Some analysts warn that without retaining more STEM graduates, the United States would suffer a loss of entrepreneurship, would decline in the knowledge economy, and would lose its premier place in the world of innovation. Proponents of STEM visas cite the substantial contributions of foreign-born residents in the United States to international patent creation (25.6% in 2006). "If we don't keep the skilled people in this country after they are educated in our universities and our institutions," U.S. Chamber of Commerce President Tom Donohue said, "companies have a simple choice: If we can't get them here and they go somewhere else, we send the work to where they are." The perspective of many in the business community is that skilled immigrants make jobs rather than take jobs.[54]

Some researchers maintain that creating a STEM visa would not be prudent at this time, given the soft labor market for some STEM occupations. To support this argument, they point out that the unemployment rate of 5.2% for computer and mathematical occupations was greater than the 4.7% unemployment rate of college graduates generally in 2010, and that the unemployment rate of chemists and material scientists reached its highest level in 40 years in 2011 (6.1%).[55] Earlier research from 2007 found that STEM shortages were either overstated or misunderstood. One scholar testified about research indicating that some employers already use high-skilled temporary visa categories to recruit foreign workers at below market rates. This researcher also argued that employers have undue control over the temporary foreign workers because their legal status hinges on their employment. One study found that universities in the United States actually graduate many more STEM students than are hired each year. In 2011, a policy researcher testified that the science and engineering labor force has a substantial supply and that it is best characterized as a loose (not tight) labor market.[56]

Recently, the debate over U.S. labor market needs for STEMs has centered on occupations in computer sciences. Microsoft published a report projecting the graduation rates in computer science will not keep pace with estimated job openings through 2020 in the computer science occupations that require at least a bachelor's degree.[57] The Economic Policy Institute published a report critiquing the Microsoft analysis and arguing that there is no shortage of workers in computer-related occupations.[58]

Defining/Refining STEM Fields

As noted at the onset, there is no consensus on the definition of STEM fields within academia or federal agencies. Concerns have been expressed that linking LPR status to a STEM degree without a clearly accepted definition might well have unintended consequences. For example, as a result of revisions in federal Standard Occupational Classification (SOC) in STEM occupations, there was a significant reclassification of employees. Roughly 5% of jobs in the U.S. labor force currently are considered STEM positions. The broader the STEM fields are defined, the more likely that no critical discipline would be omitted; however, a broad definition, some maintain, would also be more susceptible to gaming and abuses.[59]

Others point out that even the narrowest definition would include widely heterogeneous disciplines with widely varied labor market conditions. For example, the engineering profession, which has evolved and splintered into sub-specialties over time, was defined in 1993 as follows:

> Civil engineers had more specialized training as structural engineers, dam engineers, water- power engineers, bridge engineers; mechanical engineers as machine-design engineers, industrial engineers, motive-

> power engineers; electrical engineers as power and communication engineers (and the latter divided eventually into telegraph, telephone, radio, television, and radar engineers, whereas the power engineers divided into fossil-fuel and nuclear engineers); mining engineers as metallic-ore mining engineers and fossil-fuel mining engineers (the latter divided into coal and petroleum engineers).[60]

Today, engineering also encompasses such professions as computer engineers designing micro- chips that use light pulses and biomedical engineers researching the structure of human cells to foster tissue growth. To deal with the issue of defining STEM fields, one witness during a 2011 House hearing recommended first clarifying the policy motivations for the immigration benefit and then assigning the task of enumerating eligible fields to an agency or inter-agency work- group as an option.[61]

Confounding the discussion of what disciplines should be considered "STEM fields" is the perspective that STEM degrees are not as important as competency in a core set of STEM-based capabilities (referred to as STEM competency). There is an increasing demand for STEM competency in non-STEM occupations, which enables many people with STEM competencies to choose higher-paying career options in other fields. One report finds that out of every 100 students with a bachelor's degree, 19 graduate with a STEM degree but only eight are working in STEM occupations 10 years after graduation.[62]

Colleges and Universities as Immigration Gatekeepers

Proponents of STEM visa legislation assert that completion of a graduate degree in a STEM field at a U.S. institution is an excellent criterion for selecting LPRs. They argue that the United States should not be sending back the world's brightest students after educating them. They maintain that current policy not only fosters a reverse brain drain from our economy, but it also equips nations who compete with the United States in the global economy with graduates trained by our world-class universities. A witness at a 2011 hearing on STEM visas testified that a significant majority of foreign students—85% of Indian and Chinese students and 72% of European students—reported concerns about obtaining work visas. The witness warned that the United States was falling behind in the global competition for talent.[63]

Other policymakers question the potential consequences of linking LPR status to student visas. For example, the Chairman of the House Committee on the Judiciary warned that establishment of STEM visas would create an incentive for some schools to recruit tuition-paying foreign students with the lure of LPR status upon graduation and cited reports from Australia, where some institutions of higher education were deemed to be "selling education for visas."

Another study of the unintended consequences of Australia's linkage of its skilled migration program with international students reported " the failure of some former overseas students to achieve employment outcomes that were commensurate with their qualifications; and failure to obtain skill levels that would meet Australia's skill needs." Further concerns are the possibilities that the incentive of STEM visas may lead to foreign nationals displacing qualified American students or may lead to an over-production of advanced STEM degrees.

Some experts advise that institutional eligibility should be carefully defined to prevent "diploma mills" from cashing in on the STEM visas and offer options such as limiting the eligible institutions to those requiring residency at the institution or receiving research funding from NSF or the National Institutes of Health.[64]

Context of Broader Immigration Reform

The challenge in formulating an immigration system is structuring it to represent the country's values, priorities, and needs. The difficulty is designing a system that operationalizes these specific values, priorities, and needs into immigration policy.

STEM visas are appealing to some because they offer an option that appears to be solely merit-based. Yet, such specification of merit itself provokes debate over what human capital, personal traits, and values prospective immigrants should bring to a country. Some maintain that the STEM visas legislation has broad enough support to be enacted without tackling the thornier issues that confound comprehensive immigration reform, such as the immigration status of unauthorized students who were brought to the United States as children by their parents.[65] Others maintain it merely begs the question by increasing legal immigration without addressing wider reforms.

APPENDIX. OTHER HIGH-SKILLED TEMPORARY EMPLOYMENT CATEGORIES

Cultural Exchange Workers: J Visas[66]

The broadest category for cultural exchange is the J visa, which includes professors and research scholars, students, foreign medical graduates, teachers, resort workers, camp counselors and au pairs who are participating in an approved exchange visitor program. The U.S. Department of State's Bureau of Educational and Cultural Affairs (BECA) is responsible for approving the cultural exchange programs. J visa holders are admitted for the period of the program. Many foreign nationals on J-1 visas are permitted to work as part of their cultural exchange program participation. The J cultural exchange visas have expanded over time from visas issued for educational, research, or scholarship purposes to visas issued for programs engaged in more mundane tasks, such as child care, resort work, or camp counseling. Today, the J visas may be issued to over a dozen subcategories of exchange visitors. Many of the J visa holders are coming to work in the United States, and some may be employed in STEM fields.

Multinational Executive and Specialist Employees: L Visas

Intracompany transferees who are executive, managerial, or have specialized knowledge and who are employed with an international firm or corporation are admitted on the L visas. The prospective L nonimmigrant must demonstrate that he or she meets the qualifications for the particular job as well as the visa category. The foreign national must have been employed by the firm for at least six months in the preceding three years in the capacity for which the transfer is sought. The alien must be employed in an executive capacity, a managerial capacity, or have specialized knowledge of the firm's product to be eligible for the L visa. STEM graduates are most likely to fulfill the specialized knowledge element. The INA does not require firms who wish to bring L intracompany transfers into the United States to demonstrate that U.S. workers will not be adversely affected in order to obtain a visa for the transferring employee.

International Investors: E Visas

Aliens who are treaty traders enter on E-1 visas, whereas those who are treaty investors enter on E-2 visas. An E-1 treaty trader visa allows a foreign national to enter the United States for the purpose of conducting "substantial trade" between the United States and the country of which the person is a citizen. An E-2 treaty investor can be any person who comes to the United States to develop and direct the operations of an enterprise in which he or she has invested, or is in the process of investing, a "substantial amount of capital." Entrepreneurial STEM graduates, in particular, may utilize the E-2 visa. Both these E-class visas require that a treaty exist between the United States and the principal foreign national's country of citizenship.[67]

Persons with Outstanding and Extraordinary Ability: O Visas

Persons with extraordinary ability in the sciences, arts, education, business, or athletics can be admitted on O visas. Generally, the O visa is reserved for the highest level of accomplishment and covers a fairly broad set of occupations and endeavors, including athletics and entertainment. Regulations implementing the O-1 visa define extraordinary ability in the field of science, education, business, or athletics as "a level of expertise indicating that the person is one of the small percentage who have arisen to the very top of the field of endeavor."[68] In FY2010, the State Department issued 8,589 O-1 visas.

End Notes

[1] For a fuller discussion of science, technology, engineering, or mathematics (STEM) fields, see CRS Report R42642, *Science, Technology, Engineering, and Mathematics (STEM) Education: A Primer*, by Heather B. Gonzalez and Jeffrey J. Kuenzi; CRS Report R42430, *America COMPETES 2010 and the FY2013 Budget*, by Heather B. Gonzalez; and CRS Report R42470, *An Analysis of STEM Education Funding at the NSF: Trends and Policy Discussion*, by Heather B. Gonzalez.

[2] For a complete discussion of the employment-based visa queue, see CRS Report R42048, *Numerical Limits on Employment-Based Immigration: Analysis of the Per-Country Ceilings*, by Ruth Ellen Wasem.

[3] The other two economic preference categories are: special immigrants who largely consist of religious workers and certain former employees of the U.S. government; and investors who invest at least $1 million (or less money in rural areas or areas of high unemployment) to create at least 10 new jobs. §203(b) of INA; 8 U.S.C. §1153.

[4] Committee on Science, Engineering, and Public Policy, *Policy Implications of International Graduate Students and Postdoctoral Scholars in the United States*, National Academies Press, 2005; Vivek Wadhwa, Guillermina Jasso, and Ben Rissing, et al., *Intellectual Property, the Immigration Backlog, and a Reverse Brain-Drain, part III*, Duke University, New York University, Harvard Law School and the Ewing Marion Kauffman Foundation, August 2007; and, David Langdon, George McKittrick, and David Beede, et al., *STEM: Good Jobs Now and for the Future*, U.S. Department of Commerce Economics and Statistics Administration, 03-11, July 2011, http://www.esa.doc.gov/sites/ default/files/ reports/documents/stemfinalyjuly14_1.pdf.

[5] Thomas Lemieux, "The Changing Nature of Wage Inequality," *Journal of Population Economics*, vol. 21, no. 1 (January 2008); David Card and John E. DiNardo, "Skill-Biased Technological Change and Rising Wage Inequality: Some Problems and Puzzles," *Journal of Labor Economic*, vol. 20, no. 4 (2002); B. Lindsay Lowell, Hal Salzman, and Hamutal Bernstein, *Steady as She Goes? Three Generations of Students through the Science and Engineering Pipeline*, Rutgers University and the Institute for the Study of International Migration, October 2009; and, U.S. Congress, House Committee on the Judiciary, Subcommittee on Immigration Policy and Enforcement, *STEM the Tide: Should America Try to Prevent an Exodus of Foreign Graduates of U.S. Universities with Advanced Science Degrees,* testimony of Dr. B. Lindsey Lowell, 112th Cong., 1st sess., October 5, 2011; and David Langdon, George McKittrick, and David Beede, et al., *STEM: Good Jobs Now and for the Future*, U.S. Department of Commerce Economics and Statistics Administration, 03-11, July 2011, http://www.esa.doc.gov/sites/default/files/reports/documents/ stemfinalyjuly14_1.pdf.

[6] Peter Einaudi, *Two Decades of Increasing Diversity More than Doubled the Number of Minority Graduate Students in Science and Engineering*, National Science Foundation, NSF 11-319, July 2011, http://www.nsf.gov/statistics/infbrief/ nsf11319/nsf11319.pdf.

[7] For a detailed list of the Immigration and Custom Enforcement STEM-designated degree programs, see http://www.ice.gov/doclib/sevis/pdf/stem-list-2011.pdf.

[8] National Center for Education Statistics, *Postsecondary Awards in Science, Technology, Engineering, and Mathematics, by State: 2001 and 2009*, U.S. Department of Education, April 2011, http://nces.ed.gov/pubs2011/ 2011226.pdf.

[9] CRS Report RL32188, *Monitoring Foreign Students in the United States: The Student and Exchange Visitor Information System (SEVIS)*, by Alison Siskin; and CRS Report RL31146, *Foreign Students in the United States: Policies and Legislation*, by Chad C. Haddal.

[10] Nonimmigrants are admitted for a designated period of time and a specific purpose.There are 24 major nonimmigrant visa categories, which are commonly referred to by the letter and numeral that denotes their subsection in the INA; for example, B-2 tourists, E-2 treaty investors, F-1 foreign students, or H-1B temporary professional workers. CRS Report RL31381, *U.S. Immigration Policy on Temporary Admissions*, by Ruth Ellen Wasem.

[11] Those students who wish to pursue a non-academic (e.g., vocational) course of study apply for an M visa. Their spouses and children may accompany them as M-2 nonimmigrants. Some graduate students use the J visa for cultural exchange, which covers educational, research or scholarship purposes.

[12] 8 C.F.R. 214.2(f)(15)(i).

[13] SEVIS is the foreign student monitoring database. For background on SEVIS, see CRS Report RL32188, *Monitoring Foreign Students in the United States: The Student and Exchange Visitor Information System (SEVIS)*, by Alison Siskin.

[14] Until the 2000s, the enrollment of U.S. citizens in graduate science and engineering programs had not kept pace with that of foreign students in these programs. NSF researchers cite the growth in the numbers of U.S. citizens and permanent residents pursuing graduate-level study in STEM fields in recent years as the reason the percentage of foreign STEM students has remained level. Peter Einaudi, *Two Decades of Increasing Diversity More than Doubled the Number of Minority Graduate Students in Science and Engineering*, National Science

Foundation, NSF 11-319, July 2011, http://www.nsf.gov/statistics/infbriefnsf11319/ nsf11319.pdf.

[15] To access the IPEDS data, go to http://nces.ed.gov/ipeds/ on the National Center for Education Statistics website.

[16] Although based upon data from ICE, the STEM fields presented in **Figure 2**, **Table 1**, and **Table 2** use the broader NSF definition of STEM. Joan Burrelli, *Foreign Science and Engineering Students in the United States,* National Science Foundation, NSF 10-324, July 2010, http://www.nsf.gov/statistics/infbrief/nsf10324/.

[17] E-Verify is an electronic employment eligibility verification program that U.S. employers voluntarily use to confirm the new hires' employment authorization through Social Security Administration and, if necessary, DHS databases. CRS Report R40446, *Electronic Employment Eligibility Verification*, by Andorra Bruno.

[18] 8 C.F.R. 214.2(f)(10).

[19] Attestation was part of a compromise package on H-1B visas that included annual numerical limits in the Immigration Act of 1990 (P.L. 101-649). See CRS Report RL30498, *Immigration: Legislative Issues on Nonimmigrant Professional Specialty (H-1B) Workers*, by Ruth Ellen Wasem.

[20] INA §212(n); 8 C.F.R. §214.2(h)(4). For a further discussion of labor attestations, see CRS Report RL30498, *Immigration: Legislative Issues on Nonimmigrant Professional Specialty (H-1B) Workers*, by Ruth Ellen Wasem.

[21] U.S. Citizenship and Immigration Services, *Characteristics of H-1B Specialty Occupation Workers, Fiscal Year 2010 Annual Report*, Department of Homeland Security, August 4, 2011.

[22] Ibid.

[23] §501 of P.L. 109-13, the Emergency Supplemental Appropriations Act for Defense, the Global War on Terror, and Tsunami Relief, 2005.

[24] U.S. Citizenship and Immigration Services, *Extension of Post-Completion Optional Practical Training (OPT) and F- 1 Status for Eligible Students under the H-1B Cap-Gap Regulations*, U.S. Department of Homeland Security, April 2, 2010.

[25] Not all companies, however, seek to convert H-1B employees to LPR status. Research by Professor Ron Hira of the Rochester Institute of Technology indicates that many of the largest users of the H-1B visa sponsor few, if any, of their H-1Bs for permanent residency. U.S. Congress, Senate Committee on the Judiciary, Subcommittee on Immigration, Refugees and Border Security, *The Economic Imperative for Enacting Immigration Reform*, 112th Cong., 1st sess., July 26, 2011.

[26] The other categories permitted dual intent are intracompany transfers employed with international firms who enter on L visas and foreign nationals with V visas for family-related nonimmigrant. §214(b) of the INA; 8 U.S.C. §1184(b).

[27] Numbers not used in one category roll down to the next category. CRS Report RL32235, *U.S. Immigration Policy on Permanent Admissions*, by Ruth Ellen Wasem.

[28] For examples of legislation in the 112th Congress that would amend the fifth preference to foster STEM graduates who are entrepreneurs, see H.R. 1114 and S. 565.

[29] Randall Monger and James Yankay, *U.S. Legal Permanent Residents: 2010*, U.S. Department of Homeland Security Office of Immigration Statistics, Annual Flow Report, March 2011, http://www.dhs.gov/xlibrary/assets/statistics/ publications/lpr_fr_2010.pdf.

[30] See CRS Report RL33977, *Immigration of Foreign Workers: Labor Market Tests and Protections*, by Ruth Ellen Wasem.

[31] CRS analyzed the U.S. Department of Homeland Security Office of Immigration Statistics LPR data for fiscal years 2000 through 2009. There were a total of 10.3 million LPRs over the decade, of whom 69% reported an occupation. The occupation data analyzed in this report are the subset of 676,642 principal immigrants who received employment 1st, 2nd, or 3rd preference LPR status from fiscal years 2000 through 2009.

[32] A substantial visa retrogression occurred in July 2007. The *Visa Bulletin for July 2007* listed the visa priority dates as "current" for the employment-based preferences (except for the unskilled other worker category). On July 2, 2007, however, the State Department issued an *Update to July Visa Availability* that retrogressed the dates to "unavailable." The State Department offered the following explanation: "The sudden backlog reduction efforts by Citizenship and Immigration Services Offices during the past month have resulted in the use of almost 60,000 Employment numbers.... Effective Monday July 2, 2007 there will be no further authorizations in response to requests for Employment-based preference cases." U.S. Department of State, Bureau of Consular Affairs, *Visa Bulletin,* No. 108, is available at http://travel.state.gov/visa/frvi/bulletin/bulletin_3266.html.

[33] U.S. Congress, House Committee on the Judiciary, Subcommittee on Immigration Policy and Enforcement, *H-1B Visas: Designing a Program To Meet the Needs of the U.S. Economy and U.S. Workers*, 112th Cong., 1st sess., March 31, 2011; and U.S. Congress, House Committee on the Judiciary, Subcommittee on Immigration Policy and Enforcement, *STEM the Tide: Should America Try to Prevent an Exodus of Foreign Graduates of U.S. Universities with Advanced Science Degrees*, 112th Cong., 1st sess., October 5, 2011.

[34] U.S. Congress, Senate Committee on the Judiciary, Subcommittee on Immigration, Refugees and Border Security, *The Economic Imperative for Enacting Immigration Reform*, 112th Cong., 1st sess., July 26, 2011

[35] For a discussion of H.R. 3012, which would address high-skilled immigration by lifting the employment-based country ceilings, see CRS Report R42048, *Numerical Limits on Employment-Based Immigration: Analysis of the Per- Country Ceilings*, by Ruth Ellen Wasem.

[36] Dual intent refers to exceptions to §214(b) of the INA that permit foreign nationals to have pending petitions for both nonimmigrant and immigrant visas.

[37] For information on similar fees, see CRS Report RL31973, *Programs Funded by the H-1B Visa Education and Training Fee, and Labor Market Conditions for Information Technology (IT) Workers*, by Linda Levine and Blake Alan Naughton.

[38] Not to be confused with the StartUp Visa Act of 2011 ((H.R. 1114/S. 565) which would amend the INA to establish an employment-based, conditional immigrant visa (StartUp visa) for a sponsored alien entrepreneur: (1) with required amounts of financial backing from a qualifying investor, government entity, or venture capitalist; and (2) whose commercial activities will generate required levels of employment, revenue, or capital investment.

[39] For background on the Diversity Visa Lottery, see CRS Report R41747, *Diversity Immigrant Visa Lottery Issues*, by Ruth Ellen Wasem.

[40] The STEM Jobs Act of 2012 is posted on the website of the House Rules Committee at http://docs.house.gov/.

[41] Roll Call 590, available at http://clerk.house.gov/evs/2012/roll590.xml.

[42] For further background on the "V" visa, see CRS Report RL31381, *U.S. Immigration Policy on Temporary Admissions*, by Ruth Ellen Wasem, p.8.

[43] For a discussion of the family-based backlogs, see CRS Report RL32235, *U.S. Immigration Policy on Permanent Admissions*, by Ruth Ellen Wasem, pp. 11-13.

[44] Roll Call 613, available at http://clerk.house.gov/evs/2012/roll613.xml.

[45] For a full discussion of the data problems in estimating backlogs of approved petitions, see CRS Report R42048, *Numerical Limits on Employment-Based Immigration: Analysis of the Per-Country Ceilings*, by Ruth Ellen Wasem.

[46] The study attempted to address the unknown cases pending who were waiting in the queue for a numerically limited visa and estimated that an additional 150,000 prospective LPRs from India were waiting in the professional and skilled worker preference queue. That estimate yielded a total of 210,000 Indians in the third preference caseload. Stuart Anderson, *Waiting and More Waiting: America's Family and Employment-based Immigration System,* National Foundation for American Policy, NFAP Policy Brief, October 2011. (Hereinafter Anderson,

Waiting and More Waiting: America's Family and Employment-based Immigration System, NFAP Policy Brief, October 2011.)

[47] The study assumed that 55,000 Chinese were already in the third preference caseload. Anderson, *Waiting and More Waiting: America's Family and Employment-based Immigration System,* NFAP Policy Brief, October 2011.

[48] Anderson, *Waiting and More Waiting: America's Family and Employment-based Immigration System,* NFAP Policy Brief, October 2011.

[49] Anderson, *Waiting and More Waiting: America's Family and Employment-based Immigration System,* NFAP Policy Brief, October 2011.

[50] Anderson, *Waiting and More Waiting: America's Family and Employment-based Immigration System,* NFAP Policy Brief, October 2011.Anderson also factors in the calculation of removing the per-country ceilings on employment- based immigrants. For more on this option, see CRS Report R42048, *Numerical Limits on Employment-Based Immigration: Analysis of the Per-Country Ceilings*, by Ruth Ellen Wasem.

[51] Michael G. Finn, *Stay Rates of Foreign Doctorate Recipients from U.S. Universities, 2009*, Division of Science Resources Statistics of the National Science Foundation by Oak Ridge Institute for Science and Education through an interagency agreement with Department of Energy, January 2012, http://orise.orau.gov/files/sep/stay-rates-foreign- doctorate-recipients-2009.pdf. (Hereinafter Finn, *Stay Rates of Foreign Doctorate Recipients from U.S. Universities, 2009*.)

[52] Finn, *Stay Rates of Foreign Doctorate Recipients from U.S. Universities, 2009.*

[53] "While stay rates of India and China decline with years since graduation, this decline has been very slight and smaller than the corresponding decline in stay rates observed by all other countries combined." Michael G. Finn, *Stay Rates of Foreign Doctorate Recipients from U.S. Universities, 2009*, Division of Science Resources Statistics of the National Science Foundation by Oak Ridge Institute for Science and Education through an interagency agreement with Department of Energy, January 2012, http://orise.orau.gov/ files/sep/stay-rates-foreign-doctorate-recipients-2009.pdf.

[54] National Academy of Sciences, National Academy of Engineering, and Institute of Medicine, *Rising Above the Gathering Storm: Energizing and Employing America for a Brighter Economic Future*, Committee on Prospering in the Global Economy of the 21st Century, 2007; Brookings Institution and George Mason University, *Immigration Policy: Highly Skilled Workers and U.S. Competitiveness and Innovation*, Forum hosted by the Brookings Center for Technology Innovation and the George Mason Center for Science and Technology Policy, February 7, 2011; Vivek Wadhwa, Guillermina Jasso, and Ben Rissing, et al., *Intellectual Property, the Immigration Backlog, and a Reverse Brain-Drain, part III*, Duke University, New York University, Harvard Law School and the Ewing Marion Kauffman Foundation, August 2007; U.S. Congress, House Committee on the Judiciary, Subcommittee on Immigration Policy and Enforcement, *H-1B Visas: Designing a Program To Meet the Needs of the U.S. Economy and U.S. Workers*, 112th Cong., 1st sess., March 31, 2011; U.S. Congress, House Committee on the Judiciary, Subcommittee on Immigration Policy and Enforcement, *STEM the Tide: Should America Try to Prevent an Exodus of Foreign Graduates of U.S. Universities with Advanced Science Degrees?,* 112th Cong., 1st sess., October 5, 2011; and, Mallie Jane Kim, "Chamber of Commerce, Bloomberg Push Immigration Reform," *U.S. News & World Report*, September 28, 2011.

[55] U.S. Congress, Senate Committee on the Judiciary, Subcommittee on Immigration, Refugees and Border Security, *The Economic Imperative for Enacting Immigration Reform*, answers to questions for the record, witness Professor Ron Hira, 112th Cong., 1st sess., July 26, 2011; and Rudy M. Baum, "Unemployment Data Worst In 40 Years," *Chemical and Engineering News*, March 21, 2012.

[56] Richard Freeman, "The Market for Scientists and Engineers," *NBER Reporter*, no. 3 (Summer 2007); Rudy M. Baum, "Unemployment Data Worst In 40 Years," *Chemical and Engineering News*, March 21, 2012; U.S. Congress, Senate Committee on the Judiciary,

Subcommittee on Immigration, Refugees and Border Security, *The Economic Imperative for Enacting Immigration Reform*, answers to questions for the record, witness Professor Ron Hira, 112th Cong., 1st sess., July 26, 2011U.S. Congress, House Committee on the Judiciary, Subcommittee on Immigration Policy and Enforcement, *H-1B Visas: Designing a Program To Meet the Needs of the U.S. Economy and U.S. Workers*, testimony of Professor Ron Hira, 112th Cong., 1st sess., March 31, 2011; and, U.S. Congress, House Committee on the Judiciary, Subcommittee on Immigration Policy and Enforcement, *STEM the Tide: Should America Try to Prevent an Exodus of Foreign Graduates of U.S. Universities with Advanced Science Degrees,* testimony of Dr. B. Lindsey Lowell, 112th Cong., 1st sess., October 5, 2011.

[57] Microsoft Corporation, *A National Talent Strategy" Ideas for Securing U.S. Competitiveness and Economic Growth* (2012), http://www.microsoft.com/en-us/news/download/presskits/citizenship/MSNTS.pdf.

[58] Daniel Costa, *STEM labor shortages?: Microsoft Report Distorts Reality About Computing Occupations*, Economic Policy Institute, November 19, 2012, http://www.epi.org/publication/pm195-stem-labor-shortages-microsoft-report- distorts/.

[59] "The large decline in the numbers of people who used to be grouped in this field is that the definition of "computer scientists and systems analysts" has become much more precise, and many of the people who might have been included in the more broadly defined profession during the 1990s are now grouped elsewhere under numerous new occupational titles for specialized types of information technology people, such as software engineers, network analysts, and database administrators." Richard Ellis, *Effects of Recent Revisions in Federal Standard Occupational Classification (SOC) Categories on Counts of the Employment of STEM Professionals*, Commission on Professionals in Science and Technology, May 2, 2007, http://www.cpst.org/NewPubs.cfm; Robert D. Atkinson, "Why the Current Education Reform Strategy Won't Work," *Issues in Science and Technology*, Spring 2012, pp. 29-36; and U.S. Congress, House Committee on the Judiciary, Subcommittee on Immigration Policy and Enforcement, *STEM the Tide: Should America Try to Prevent an Exodus of Foreign Graduates of U.S. Universities with Advanced Science Degrees?*, testimony of Barmack Nassirian, 112th Cong., 1st sess., October 5, 2011.

[60] Joseph W. Barker, *McGraw-Hill Encyclopedia of Engineering*, ed. McGraw-Hill, 2nd ed. (1993), pp. 409-410.

[61] U.S. Congress, House Committee on the Judiciary, Subcommittee on Immigration Policy and Enforcement, *STEM the Tide: Should America Try to Prevent an Exodus of Foreign Graduates of U.S. Universities with Advanced Science Degrees?*, testimony of Barmack Nassirian, 112th Cong., 1st sess., October 5, 2011.

[62] "At the graduate degree level of educational attainment, it is clear that both Managerial and Professional occupations and Healthcare Professional occupations pay substantially more than STEM." Anthony P. Carnevale, Nicole Smith, and Michelle Melton, *STEM: Science, Technology, Engineering, Mathematics*, Georgetown University Center on Education and the Workforce, October 2011, V=http://cew.georgetown.edu/stem/.

[63] U.S. Congress, Joint Economic Committee, *Fueling Local Economies: Research, Innovation and Jobs*, testimony of Robert Litan, 111th Cong., June 29, 2010; U.S. Congress, House Committee on the Judiciary, Subcommittee on Immigration Policy and Enforcement, *STEM the Tide: Should America Try to Prevent an Exodus of Foreign Graduates of U.S. Universities with Advanced Science Degrees?*, testimony of Vivek Wadhwa, 112th Cong., 1st sess., October 5, 2011.

[64] U.S. Congress, House Committee on the Judiciary, Subcommittee on Immigration Policy and Enforcement, *STEM the Tide: Should America Try to Prevent an Exodus of Foreign Graduates of U.S. Universities with Advanced Science Degrees?*, 112th Cong., 1st sess., October 5, 2011; Elsa Koleth, *Overseas Students: Immigration Policy Changes 1997– May 2010*, Parliament of Australia, Parliamentary Library Background Note, June 18, 2010, http://www.aph.gov.au/ binaries/library/pubs/bn/sp/overseasstudents.pdf.

[65] For a complete analysis of these immigration issues, see CRS Report R41704, *Overview of Immigration Issues in the 112th Congress*, by Ruth Ellen Wasem; and CRS Report R42036, *Immigration Legislation and Issues in the 112th Congress*, coordinated by Andorra Bruno.

[66] The Q visa is also an employment-oriented cultural exchange program. Its stated purpose is to provide practical training and employment as well as to share the history, culture, and traditions of the United States and of the exchange visitor's home country. While STEM graduates are not typically recruited in Q exchange programs, they are not precluded.

[67] See CRS Report RL33844, *Foreign Investor Visas: Policies and Issues.*

[68] 8 C.F.R. '214.2(o)(3)(ii).

In: The U.S. Science and Engineering Workforce ISBN: 978-1-62948-932-2
Editors: Russell B. Hartley

Chapter 3

STEM EDUCATION: PREPARING FOR THE JOBS OF THE FUTURE*

Staff of Joint Economic Committee Chairman, Senator Bob Casey

THE VALUE OF STEM SKILLS IN THE 21ST CENTURY ECONOMY

The United States has traditionally produced the world's top research scientists and engineers, leading to breakthrough advances in science and technology. This technological innovation has been a primary driver of U.S. economic growth, with studies showing that half or more of economic growth in the United States over the past fifty years is attributable to improved productivity resulting from innovation.[1] The late 20th century ushered in vast improvements in computer and information technologies, as well as biomedical technologies. These innovations are changing the way we live, work, and play in marvelous and unforeseen ways. Fully capturing the economic benefits of existing and undiscovered technologies will require a steady stream of Americans equipped with science, technology, engineering, and math (STEM) knowledge, skills, and abilities.

---

* This is an edited, reformatted and augmented version of a Report by the Joint Economic Committee Chairman's Staff Senator Bob Casey, Chairman, dated April, 2012.

The need for workers with STEM skills is heightened in today's global economy. Technological innovation improves the competitive position of U.S. industries, drives export growth, and supports high-quality jobs.[2] Additionally, demand for STEM-capable workers has increased even in traditionally non-STEM fields due to the diffusion of technology across industries and occupations.[3] Technological progress can improve workers' quality of life by improving working conditions and increasing wages. However, for workers to benefit from the economic gains generated by innovation, they must have the knowledge and skills needed to work effectively in jobs requiring STEM competencies.[4] Improving access to quality STEM education will strengthen the caliber of the U.S. workforce, drive economic growth, and keep the U.S. competitive.

WHAT IS STEM?

The acronym STEM stands for science, technology, engineering, and math. However, there is some disagreement with regard to precisely which degrees and occupations fall into the category of STEM. In general, for the purposes of this report, we consider STEM to include life sciences (except medical sciences), physical sciences, mathematics and statistics, computing, and engineering. We do not include social sciences in our definition of STEM fields. However, sources referenced in this report have differing definitions of STEM. For example, the Department of Commerce includes certain STEM-related managerial occupations in its definition of STEM, and the OECD includes manufacturing and processing, and architecture and building. These differences in definitions across sources can complicate comparisons or analyses of trends in STEM.

DEMAND FOR QUALIFIED STEM WORKERS IS PROJECTED TO INCREASE IN THE FUTURE

While it is difficult to project trends in the labor market, the demand for STEM-skilled workers is expected to continue to increase in the future, as both the number and proportion of STEM jobs are projected to grow.[5] New Bureau of Labor Statistics data show that employment in STEM occupations is expected to expand faster than employment in non-STEM occupations from 2010 to 2020 (by 17 versus 14 percent).[6] While the projected difference between the STEM and non-STEM employment growth rates does not appear to be particularly dramatic, this small difference is due in large part to the Great Recession and the severe job losses concentrated in non-STEM occupations. (See **Figure 1**.) For example, while jobs in both construction and extraction and computer and mathematical occupations are expected to grow by over 20 percent over the course of the current decade, the employment growth in construction is a result of recovery from job losses over the past few years, while the projected growth in computing and mathematics is in addition to jobs gained in those occupations despite the recession.[7]

In addition to government projections of employment growth in STEM fields, business organizations and other groups have issued numerous reports and surveys that suggest that there is a heightening need for qualified STEM workers – both those with highly specialized skills as well as those with a more general knowledge of STEM concepts. For example, even at the height of the recession, a survey of manufacturers found that over one-third were experiencing shortages of engineers and scientists – and most of them anticipated greater shortages in the future.[8] In addition, an earlier survey found that over half of manufacturers believe that the public education system insufficiently prepares students with the math and science skills necessary to succeed in the workplace.[9] Furthermore, concerns regarding shortages of skilled workers are compounded by the pending retirements of many baby boomers, an issue cited by both private industry and government officials.[10]

High Demand for STEM Workers Leads to Lower Levels of Unemployment and Higher Wages for Those Workers

Lower unemployment rates and higher relative wages in STEM occupations indicate the strong demand for STEM workers. The relative strength of the labor market for STEM workers pre-dated the recent recession, and it persisted during and after the recession. The STEM unemployment rate peaked at 5.5 percent in 2009, while the unemployment rate in non-STEM occupations was still climbing, reaching nearly 10 percent in 2010.[11] Employment in some STEM sectors has been particularly recession-proof, including computer and mathematical occupations, where the unemployment rate was 4.1 percent in 2011 – less than half of the overall unemployment rate of 8.9 percent.[12] The life, physical, and social sciences, and architecture and engineering occupations – which predominately consist of STEM workers – also had lower unemployment rates in 2011, of 3.8 percent and 5.1 percent, respectively.[13]

Employees in STEM occupations typically have higher average salaries than other workers. In 2010, STEM workers earned 26 percent more than non-STEM workers, even after accounting for other factors that affect pay, such as age, gender, race, location, industry, and union status.[14] This so-called "STEM premium" is persistent and has grown over time – in 1994, the STEM premium was just 18 percent. While the size of this premium varies with education, it exists for all levels of education. The differential is greatest among workers with less than a bachelor's degree, who make, on average, a bit over 30 percent more per hour than non-STEM workers with a similar level of education.[15] Among more highly-educated individuals, bachelor's degree recipients make about 23 percent more, and those with a graduate degree or higher earn 12 percent more than their counterparts in non-STEM occupations.[16]

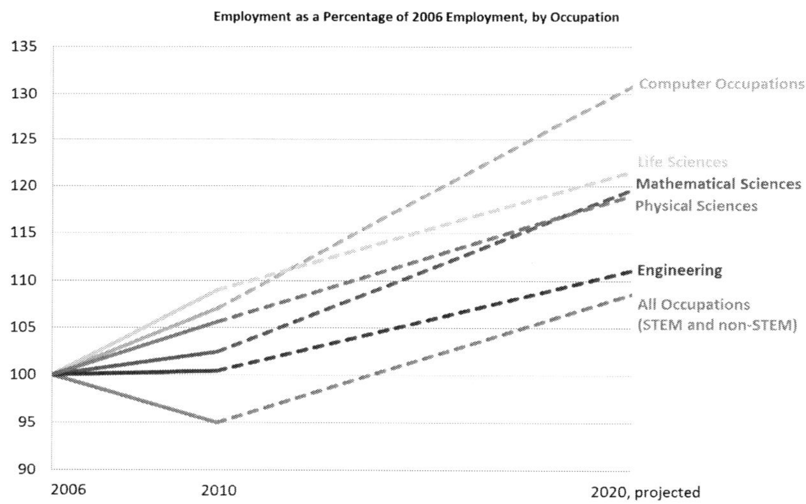

Source: Chairman's staff of the Joint Economic Committee based on data from the Bureau of Labor Statistics. The BLS does not project employment for individual years from 2010-20. For the purposes of this chart, Life Sciences excludes Medical Sciences.

Figure 1. Sustained Growth is Projected for STEM Occupations.

THE SUPPLY OF STEM TALENT IS NOT KEEPING UP WITH DEMAND

Despite the clear demand for STEM talent by domestic employers, the U.S. is failing to produce an ample supply of workers to meet the growing needs of both STEM and non-STEM employers. The existing STEM pipeline leaves too many students without access to quality STEM education, and without the interest and ability to obtain a degree or work in STEM. Current statistics on STEM education in the U.S. highlight the challenge facing educators and policymakers, making it clear that the United States must do more to build a strong STEM workforce if it is to remain competitive in the global economy.

A Smaller Percentage of Students Are Pursuing Post-Secondary STEM Degrees

While the number of students receiving degrees from four-year institutions has increased in the U.S. as access to higher education has expanded over the past several decades, the share of students graduating with STEM degrees has declined. The share of bachelor's degrees awarded in STEM fields peaked at 24 percent in 1985; by 2009, the share had fallen to 18 percent.[17] The share of master's degrees in STEM fields dropped from 18 percent to 14 percent over the same period.[18] (See **Figure 2**.) However, the core competencies obtained by STEM graduates are increasingly in demand by non-STEM employers. This has led to a growing number of STEM graduates working in non-STEM fields – while STEM employers continue to cite a shortage of qualified workers.

The share of doctorates in STEM fields was more stable between 1985 and 2009.[19] (See **Figure 3**.) However, the share of those degrees going to domestic students dropped from roughly three- fourths (74 percent) to a little more than half in 2006 (54 percent).[20] Foreign-born doctorates may take their expertise back to their home countries rather than filling the shortage of STEM workers.

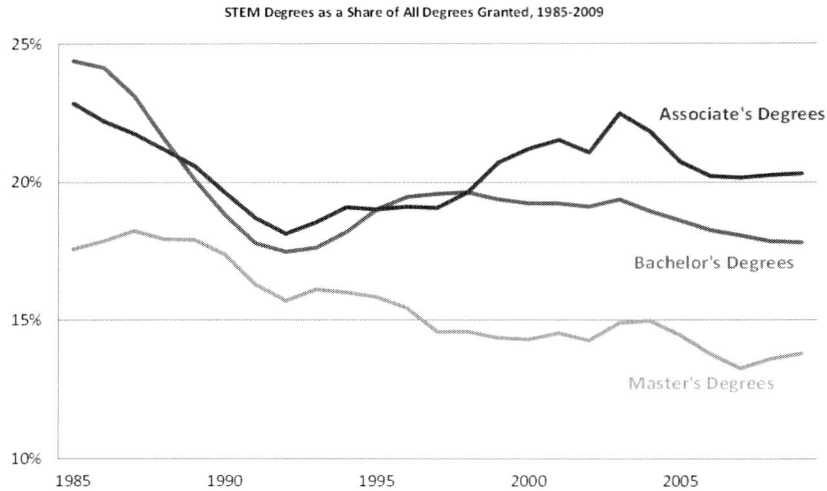

Source. Chairman's staff of the Joint Economic Committee based on data from the Department of Education's National Center for Education Statistics: Integrated Postsecondary Education Data. STEM Degrees include degrees in Engineering. Physical Sciences, Geosciences. Math and Computer Sciences, and Life Sciences (except Medical Sciences).

Figure 2. A Smaller Percentage of Degrees Are STEM Degrees.

Gender, Racial, and Ethnic Gaps Impact the Supply of STEM Workers

Of particular concern are the persistent STEM participation and achievement gaps across different demographic groups. In particular, women are less likely than men to pursue degrees in STEM, and black and Hispanic students are less likely than white students. If these disparities were corrected, the United States would be better able to fulfill the demand for STEM workers.

In 2009, women earned 57 percent of all bachelor's degrees awarded, up from 54 percent in 1993.[21] However, at the same time, the share of bachelor's degrees awarded to women in mathematics and statistics declined by 4 percent and in computer science by 10 percent.[22] Consequently, while women have comprised a growing share of the college-educated workforce,[23] their share of the STEM workforce has not increased.[24] Only 14 percent of engineers are women, as are just 27 percent of individuals working in computer science and math positions.[25] Women's increased participation in the STEM workforce is essential to alleviating the shortage of STEM workers.

There is a similar underrepresentation of Hispanic and black non-Hispanic workers in the STEM workforce. Each of these groups accounts for only six percent of STEM workers, but 14 and 11 percent of overall employment, respectively.[26] From 2000 to 2009, Hispanics as a share of the overall workforce increased by four percent, while their representation among STEM occupations increased by only one percent.[27] Non-Hispanic black workers increased as a percentage of the overall workforce by one percent over this time period, while their share of STEM workers held constant.[28] In part, this can be attributed to the lower college graduation rates among these groups and the education credentials required for many STEM jobs. However, among college graduates, Hispanics and black non-Hispanics are less likely to major in STEM fields, and, among STEM majors, individuals in these groups are less likely to ultimately end up in STEM jobs.[29]

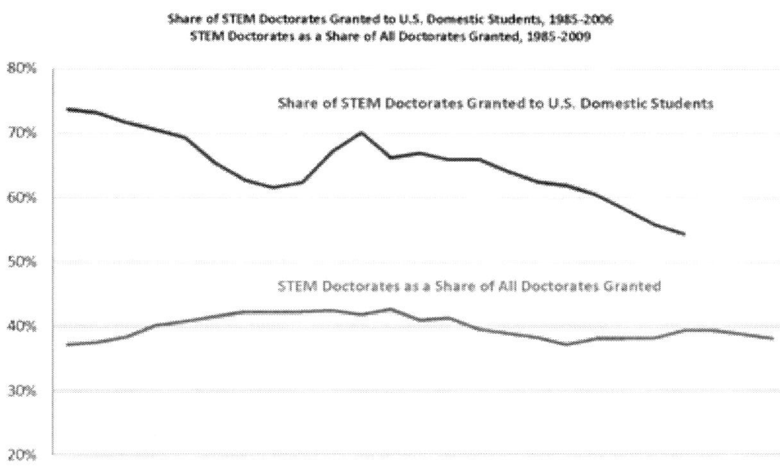

Source: Chairman's staff of the Joint Economic Committee based on data from the Department of Education's National Center for Education Statistics: Integrated Postsecondary Education Data and the National Science Foundation's Survey of Earned Doctorates. STEM Degrees include degrees in: Engineering, Physical Sciences, Geosciences, Math and Computer Sciences, and Life Sciences (except Medical Sciences).

Figure 3. A Smaller Share of STEM Doctorates Are Going to Domestic Students.

International Comparisons Suggest the U.S. Is Trailing in Developing a STEM Workforce

The United States is falling behind in international comparisons when it comes to providing an adequate supply of STEM workers. While we should be cautious in making international comparisons of countries which have dramatically varying sizes, diversity of populations, and industrial policies, the U.S. continues to fall behind key international competitors across a wide set of different measures of STEM preparedness.

While the U.S. produces by far the greatest number of STEM degree recipients among OECD countries (348,484 in 2008), adjusting for the overall number of degrees and for the population paints a different picture. By one measure – the share of students receiving degrees in STEM fields – the U.S. compares unfavorably with other global competitors. For example, NAFTA members Canada and Mexico, and many European nations including economic powerhouse Germany, graduate more STEM students as a share of all degrees than the United States does.[30] (See **Table** below.) This is also true when looking at STEM graduates as a share of the employed 25-34 year old population, where the U.S. ranks 23rd among OECD countries. American students' performance on international standardized tests suggests problems earlier in the STEM pipeline. For example, U.S. 15-year-olds rank 25th in math and 17th in science in PISA scores among OECD nations. As discussed below, other international comparisons as well as performance on domestic examinations suggest that problems in U.S. STEM education may begin as early as elementary school and continue through students' secondary and post-secondary education.

Why Are We Falling Short in STEM?

There are a number of issues with the educational pipeline in the United States that can lead to a shortage of STEM professionals. Without a strong foundation in math and science from elementary and secondary school, students may find themselves unprepared to train for and pursue careers in STEM fields. However, science and technology curriculums are often thin in K-12 education, and may not be enough to provide students with a solid foundation in STEM upon which to build.[31] Part of the problem is that it is challenging to attract and retain STEM-trained individuals to teach STEM subjects at the K-12 level when higher wages and employment opportunities outside of the education sector make working in a STEM profession an attractive alternative.[32] Furthermore, while the quality of math and science teaching is the greatest factor in improving student achievement in STEM fields, not enough K-12 math and science teachers have hands-on experience working in STEM.[33] Teachers may also lack an educational background in STEM. For example, the National Science Foundation (NSF) found that 36 percent of middle school science teachers and approximately 30 percent of middle school math teachers lack in-field training.[34]

In addition to improving the quality and quantity of STEM educators, a greater emphasis on communicating the benefits of STEM education to students is needed. College students may not have the information necessary regarding career prospects to make educated decisions in selecting a course of study.[35] Younger students may lack appropriate mentors, or there may be inadequate hands-on activities in the classroom to pique their interest. Having a mentor significantly influences students' decisions to pursue STEM degrees and occupations, especially for women and minorities.[36] Further, those with more opportunities to engage in STEM-related activities, such as science fairs, projects, and clubs, are more likely to go onto STEM careers and have accomplishments in STEM fields.[37] Finally, there is the matter of culture. While not easy to quantify, to the extent that math and science are not considered "cool" among image-conscious high school students, inevitably many talented young people will be turned off from pursuing degrees and careers in STEM fields. Women may be particularly unlikely to pursue STEM as a result of gender and cultural norms.[38]

Inadequate STEM Achievement at the K-12 Level

Results on international tests suggest that there are deficiencies in the current education system. On the most recent Trends in International Mathematics and Science Study (TIMSS), from 2007, U.S. fourth grade students ranked eleventh in math and eighth in science out of 36 countries, and U.S. eighth grade students ranked ninth in math and eleventh in science out of students in 48 countries.[39] Scores on the 2009 Programme for International Student Assessment (PISA) test were more discouraging, as U.S. 15-year-olds ranked below students in 24 other OECD countries in math and 16 other OECD countries in science.[40] The poor results on the PISA examination are particularly problematic because, as the NSF notes, the PISA tests (in contrast to the TIMSS exams) "emphasize students' abilities to apply skills and information learned in school (or from life experiences) to solve problems or make decisions they may face at work or other circumstances."[41] Moreover, the World Economic Forum ranks the United States 48th in the world in quality of mathematics and science education.[42]

STEM Education in an International Perspective

	Standardized Test Scores		STEM Graduates		
	Average PISA Math Score (15-year-olds) (2009)	Average PISA Science Score (15-year-olds) (2009)	STEM Graduates: Bachelor's Degree Equivalent (2008)	STEM Graduates as a Share of All Bachelor's Degree Equivalent Graduates (2008)	STEM Graduates per 100,000 employed 25-34 year-olds (2008)
United States	487	502	348,484	0.15	1,472
U.S. Rank in OECD[1]	T-25th	17th	1st	27th	23rd
OECD Average[2]	496	501	53,612	0.21	1,829
Remaining Top 15 OECD Countries (by nominal GDP)					
Australia	514	527	45,020	0.19	2,362
Canada	527	529	49,208	0.22	2,146
France	497	498	112,398	0.27	2,717
Germany	513	520	106,986	0.29	1,796
Italy	483	489	50,877	0.22	N/A
Japan	529	539	160,993	0.24	1,643
Korea, South	546	538	130,468	0.33	3,555
Mexico	419	416	97,410	0.25	1,085
Netherlands	526	522	17,372	0.14	1,039
Spain	483	488	51,309	0.24	1,488
Sweden	494	495	12,736	0.24	1,596
Switzerland	534	517	10,285	0.23	2,010
Turkey	445	454	50,911	0.18	1,536
United Kingdom	492	514	121,269	0.23	2,380

Sources: Chairman's staff of the Joint Economic Committee using data from the Programme on International Student Assessment (PISA), the Organisation for Economic Co-operation and Development (OECD), and the CIA World Factbook.

[1] The sources used for this table vary with regard to the number of OECD countries for which data is available. For the PISA scores, 34 countries are included. For the STEM graduates, STEM share of graduates, and STEM graduates as a share of the young workforce data, 30 countries are included.

[2] PISA provides OECD average test scores, and these are the averages listed in the table. The average number of STEM graduates was calculated by the Chairman's staff by taking the average of the 30 countries for which there is available data. The OECD average STEM share of degrees is calculated as the overall share of STEM degrees for the 30 countries (total STEM degrees as a share of total degrees). For the purposes of this chart, STEM degrees are defined as degrees in science (OECD educational category 400) and engineering, manufacturing, and construction (category 500), and bachelor's equivalent degrees are defined as Tertiary Type-A and Advanced Research Programs degrees. The OECD provides an OECD average figure for STEM graduates as a share of the young workforce, and this is the figure included above. The data for the last column are available at http://dx.doi.org/10.1787/888932460192.

Further, National Assessment of Educational Progress (NAEP) math scores have remained roughly level over the past decade, even after the No Child Left Behind Act sought to emphasize achievement as measured by standardized test scores in math and reading.[43] The 2011 NAEP math examination found that 60 percent of fourth-graders and 66 percent of eighth-graders were not proficient in math.[44] Eighteen percent and 28 percent, respectively, failed to score at even the basic level of competency.[45] In science, the most recent iteration of the NAEP examination (2009) found that 28 percent of fourth-graders, 37 percent of eighth-graders, and 40 percent of twelfth- graders did not meet basic standards of science knowledge.[46]

The racial and ethnic gaps in STEM degree recipients and STEM workers correspond to achievement gaps on the NAEP math and science examinations, with white students out- performing black and Hispanic students by large margins on the fourth, eighth, and twelfth grade tests.[47] While these gaps have shown some movement toward closing in recent years, their persistence suggests that more work needs to be done to provide a broad cross-section of American students with the basic knowledge needed to pursue STEM.[48]

Policies to Improve STEM Education

In order to reap the full gains to be had from technological innovation, the U.S. must employ a multifaceted approach to fostering an environment where innovation can flourish. Ultimately, improving STEM education and boosting the size and caliber of the STEM workforce in the United States will not be sufficient if research and development budgets are slashed or if proper incentives are not in place for businesses to invest in innovation. Thus, in considering policies to reduce deficits and debt, policymakers should take care to preserve funding for research and development, and to continue to support programs authorized under the *America COMPETES Act*. Additional policy actions could include making the research and development tax credit permanent, or passing further legislation to incentivize R&D, such as the *Life Sciences Jobs and Investment Act* (S. 1410).

Moreover, improving STEM education at the K-12 level and moving more young people into the STEM pipeline will not be enough if college is unaffordable or out of reach, or if the postsecondary education system more generally does not function effectively to take talented, STEM-interested young people and provide them with the degrees, certifications, and skills to succeed. This will require a continued commitment to maintaining Pell Grants and other federal supports for higher education, as well as a focus on career and technical education to provide young people with the opportunity to obtain STEM skills and certifications that prepare them for 21st century jobs, even if not culminating in a bachelor's or higher degree.

However, improving math and science education in our nation's elementary and secondary schools is a prerequisite to achieving the economic gains to be had from technological innovation and to improving the distribution of those gains. Unfortunately, education budgets have been under severe stress in the aftermath of the recession, with "at least 23 states [having] enacted identifiable, deep cuts in pre-kindergarten and/or K-12 spending."[49] Science may be more likely to be cut as funding becomes tight, since it is not tested under No Child Left Behind.[50] While policymakers should take care to reduce duplication and streamline programs when possible, significant cuts to federal education programs will undercut our ability to prepare our children to work and compete in the global economy.

In order to strengthen STEM education, policymakers should consider many of the recommendations included in recent reports on the topic. Just in the past two and a half years, these reports include the National Research Council's "A Framework for K-12 Science Education: Practices, Crosscutting Concepts, and Core Ideas" and "Successful K-12 STEM Education: Identifying Effective Approaches in Science, Technology, Engineering, and Mathematics," the President's Council of Advisors on Science and Technology's "Prepare and Inspire: K-12 Education in Science, Technology,

Engineering, and Math (STEM) for America's Future," and the Information Technology & Innovation Foundation's "Refueling the U.S. Economy: Fresh Approaches to Science, Technology, Engineering, and Mathematics (STEM) Education."

A Note on STEM Data Sources

There are numerous sources of data on trends in STEM education and STEM occupations. The National Center for Education Statistics' Integrated Postsecondary Data System is the principal source for information on STEM bachelor's and master's degrees, while the National Science Foundation's (NSF) Survey of Earned Doctorates provides the most reliable data on STEM doctorates. Both of these sources are compiled in the NSF's online data system, WebCaspar. The OECD also compiles data on STEM degree completions. Standardized test score data is available for the U.S.-based National Assessment of Educational Progress and the international Trends in International Mathematics and Science Study and Programme for International Student Assessment exams. The Bureau of Labor Statistics publishes biennial estimates of projected employment growth across industries, including in STEM fields, and disseminates information on current unemployment rates and median salaries. This and related information on STEM occupations is collected through the BLS/Census Bureau's Current Population Survey.

Specific legislation currently under consideration that would enhance economic competitiveness by supporting K-12 STEM education includes:

- *Innovate America Act* (S. 239) – Directs the Secretary of Education to award grants to state educational agencies in order to expand STEM secondary schools (schools with a strategic focus and expanded curricular offerings in STEM) and requires the Director of the National Science Foundation to provide awards for colleges with dramatic increases in students getting STEM degrees.
- *Effective STEM Teaching and Learning Act of 2011* (S. 463) – Creates a program to provide competitive grants to states to improve preschool through grade 12 education in STEM, including bolstering professional development for STEM teachers and materials used in the STEM curriculum. It would permit states to use up to 20 percent

of their grant for state- level STEM activities, including the development, implementation, or improvement of a comprehensive state STEM plan.

- *STEM 2 Act* (S. 619) – Directs the Secretary of Education to award competitive planning grants to states, Indian tribes or tribal organizations, nonprofit organizations, or institutions of higher education to develop effective STEM networks that coordinate STEM education efforts by (1) facilitating communication and collaboration among public and private STEM stakeholders, and (2) identifying STEM occupational skills needed in the future. It would also establish a competitive grant program for efforts to develop, implement, and evaluate STEM education training programs for teachers and administrators at the K-12 level.
- *STEM Master Teacher Corps Act of 2011* (S. 758) – Establishes a competitive matching grant program to support states and consortia of local educational agencies in creating a STEM Master Teachers Corps, which would provide specialized training and mentorship, and supplemental compensation, for select STEM teachers, with a focus on those in high-need public schools. The Corps members would then inform STEM education policy at the national, state, and local levels.
- *National STEM Education Tax Incentive for Teachers Act of 2011* (S. 1055) – Allows certain STEM teachers to claim a refundable tax credit for the undergraduate tuition expenses they incurred in obtaining their degree, up to $1,000 in any taxable year, for up to 10 years.
- *Computer Science Education Act of 2011* (S. 1614) – Authorizes grants for states to develop plans to strengthen elementary and secondary computer science education; authorizes five-year implementation grants to states to carry out plans for developing computer science standards, curriculum and assessments, improving access to underserved populations, developing and offering teacher professional development, and developing online courses; and implements an independent evaluation of state programs to inform further efforts to expand and improve computer science education.
- *Preparing Students for Success in the Global Economy Act of 2011* (S. 1675) – Allots formula grants to states and, through them, awards competitive subgrants to high-need local educational agencies to improve preschool and elementary and secondary school STEM education. It would require subgrants to be used for activities that

include: (1) recruiting, supporting, evaluating, and training STEM teachers; (2) developing and improving high- quality STEM curricula and instructional supports to improve student achievement in those subjects; and (3) integrating STEM instruction with instruction in reading, English language arts, or other academic subjects.

End Notes

[1] *See* Abramovitz, Moses. "Resource and Output Trends in the United States Since 1870." National Bureau of Economic Research. 1956. Available at nber.org/chapters/c5650.pdf; Solow, Robert. "A Contribution to the Theory of Economic Growth." *The Quarterly Journal of Economics*. 70(1). February 1956. Available at jstor.org/pss/1884513; Romer, Paul A., "Increasing Returns and Long Run Growth." Rochester Center for Economic Research. Working Paper No. 27. October 1985. Available at rcer.econ.rochester.edu/ RCERPAPERS/rcer_27.pdf; Jones, Charles. "Introduction: The Facts of Economic Growth." *Introduction to Economic Growth 2nd Edition*. 2001. Available at gdsnet.org/GDS/JonesChapter1B.pdf; Jones, Charles I. "Sources of U.S. Economic Growth in a World of Ideas." *The American Economic Review*. Vol. 92, No. 1. March 2002. Available at stanford.edu/~chadj/SourcesAER2002.pdf.

[2] Freeman, Richard B. "Does Globalization of the Scientific/Engineering Workforce Threaten U.S. Economic Leadership?" National Bureau of Economic Research. August 2006. Available at nber.org/chapters/c0207.pdf.

[3] Carnevale, Anthony P., Michelle Melton, and Nicole Smith. "STEM." Georgetown University Center on Education and the Workforce. October 2011. Available at www9.georgetown.edu /grad/gppi/hpi/cew/pdfs/stem-complete.pdf.

[4] The Organisation for Economic Co-Operation and Development. "The OECD Jobs Study: Facts, Analysis, Strategies." 1994. Available at oecd.org/dataoecd/42/51/1941679.pdf.

[5] Carnevale, Smith, and Melton, *supra* note 3.

[6] Chairman's Staff of the Joint Economic Committee. Calculations using data from the Bureau of Labor Statistics. "Employment Projections: 2010-20." Table 1.7 Occupational Employment and Job Openings Data, Projected 2010-20, and "Worker Characteristics, 2010." February 2012. Available at bls.gov/emp/. For the purposes of this calculation, STEM occupations are defined as in the U.S. Department of Commerce's Economics and Statistics Administration report, "STEM: Good Jobs Now and for the Future." ESA Issue Brief #03-11. July 2011. Available at esa.doc.gov/sites/default/files/reports/documents/ stemfinaly july14_1.pdf.

[7] *Ibid.*

[8] Deloitte Consulting LLP, The Manufacturing Institute, and Oracle Corporation. "People and Profitability: A Time for Change." 2009. Available at deloitte.com/assets/Dcom-UnitedStates/Local%20Assets/Documents/us_pip_peoplemanagementreport_100509.pdf.

[9] Deloitte Consulting LLP and The Manufacturing Institute. "2005 Skills Gap Report – A survey of the American Manufacturing Workforce." 2005. Available at doleta.gov/wired/files/ us_mfg_talent_management.pdf.

[10] Tom Donahue. U.S. Chamber of Commerce. "Statement of the U.S. Chamber of Commerce on the Reauthorization of the America COMPETES Act." Testimony before the House Committee on Science and Technology. January 20, 2010. Available at uschamber.com/ sites/default/files/testimony/100119_americacompetes.pdf; *See also* U.S. Department of Labor, Employment and Training Administration. "The STEM Workforce Challenge: The Role of the Public Workforce System in a National Solution for a Competitive Science,

Technology, Engineering, and Math Workforce." April 2007. Available at doleta.gov/youth_services/pdf/STEM_Report_4%2007.pdf; *See also* The Wharton School, University of Pennsylvania. "Wanted: A Shortage of Engineers to Keep the Aerospace Industry Flying: A Shortage of Engineers Threatens the Industry – and National Security." October 2009. Available at executiveeducation.wharton.upenn.edu/wharton-aerospace-defense-report/upload/A-Shortage-of-Engineers- Threatens-the-Industry-and-Nation.pdf.

[11] U.S. Department of Commerce, Economics and Statistics Administration. "STEM: Good Jobs Now and for the Future." ESA Issue Brief #03-11. July 2011. Available at esa.doc.gov/sites/default/files/reports/documents/stemfinalyjuly14_1.pdf. The Department of Commerce defined STEM as including professional and technical support occupations in computer science and mathematics, engineering, and life and physical sciences, as well as certain related managerial occupations.

[12] Bureau of Labor Statistics. Household Data, Annual Averages. 25. Unemployed Persons by Occupation and Sex. Available at bls.gov/cps/cpsa2011.pdf.

[13] *Ibid.*

[14] U.S. Department of Commerce. "STEM: Good Jobs Now and for the Future," *supra* note 11.

[15] *Ibid.*

[16] *Ibid.*

[17] Chairman's Staff of the Joint Economic Committee. Calculations using data from the Department of Education's National Center for Education Statistics: Integrated Postsecondary Education Data System. Available through the National Science Foundation's WebCASPAR data system at webcaspar.nsf.gov/.

[18] *Ibid.*

[19] Chairman's Staff of the Joint Economic Committee. Calculations using data from the National Science Foundation's Division of Science Resources Statistics: Survey of Earned Doctorates. Available through the National Science Foundation's WebCASPAR data system at webcaspar.nsf.gov/.

[20] *Ibid.*

[21] NCES IPEDS data, *supra* note 17.

[22] *Ibid.*

[23] From 2000 to 2009, women as a share of all college-educated workers increased from 46 to 49 percent. *See* U.S. Department of Commerce, Economics and Statistics Administration. "Women in STEM: A Gender Gap to Innovation." ESA Issue Brief #04-11. August 2011. Available at esa.doc.gov/sites/default/files/reports/documents/womeninstemagapto innovation8311.pdf.

[24] *Ibid.*

[25] *Ibid.*

[26] U.S. Department of Commerce, Economics and Statistics Administration. "Education Supports Racial and Ethnic Equality in STEM." ESA Issue Brief #05-11. September 2011. Available at esa.doc.gov/sites/default/files/reports/documents/educationsupportsracialandethnic qualityinstem_0.pdf.

[27] *Ibid.*

[28] *Ibid.*

[29] *Ibid.*

[30] Organisation for Economic Co-Operation and Development. *OECD.StatExtracts*. Graduates by Field of Education. Available at stats.oecd.org/Index.aspx.

[31] While there has been improvement in recent years, by 2012, there will still be 11 states that require fewer than three math classes to graduate high school, and, by 2013, 14 states that require fewer than three science classes. Moreover, a smaller percentage of secondary schools offered introductory computer science courses or an Advanced Placement computer science class in 2011 than in 2005; *See* Council of State Governments. "STEM Education: Bolstering Future American Competitiveness." September 2008. Available at csg.org/knowledgecenter/docs/TIA_STEM%20education.pdf; *See also* Computer Science Teachers

Association. "CSTA National Secondary Computer Science Survey: Comparison of Results from 2005, 2007, 2009, and 2011 Surveys." Available at csta.acm.org/Research/sub/Projects/ResearchFiles/CSTASurvey2011Comp.pdf.

[32] Temin, Peter. "Low Pay, Low Quality." *Education Next*. Vol. 3. No. 3. Summer 2003. Available at educationnext.org/low-pay-low-quality/.

[33] Business Higher Education Forum. "An American Imperative: Transforming the Recruitment, Retention, and Renewal of Our Nation's Mathematics and Science Teaching Workforce." 2007. Available at bhef.com/solutions/documents/AnAmericanImperative.pdf.

[34] The percentage of teachers lacking in-field training drops to seven percent for high school biology/life sciences teachers, 18 percent for high school physical sciences teachers, and 12 percent for high school math teachers; *See* National Science Foundation, National Science Board. "Science and Engineering Indicators 2012." NSB 12-01. 2012. Available at nsf.gov/statistics/seind12/pdf/seind12.pdf.

[35] Romer, Paul. "Should the Government Subsidize Supply or Demand in the Market for Scientists and Engineers?" National Bureau of Economic Research. Working Paper #7723. June 2000. Available at nber.org/papers/w7723.pdf?new_window=1.

[36] U.S. Government Accountability Office. "Higher Education: Federal Science, Technology, Engineering, and Mathematics Programs and Related Trends." Report to the Chairman, Committee on Rules, House of Representatives. GAO 06-114. October 2005. Available at gao.gov/new.items/d06114.pdf.

[37] Benbow, Camilla P., David Lubinski, James H. Steiger, and Jonathan Wai. "Accomplishment in Science, Technology, Engineering, and Mathematics (STEM) and Its Relation to STEM Educational Dose: A 25-Year Longitudinal Study." *Journal of Educational Psychology*. Vol. 102, No. 4. 2010. Available at vanderbilt.edu/peabody/smpy/Wai_2010_accomplishment.pdf.

[38] Park, Lora E., Rebecca T. Pinkus, Jordan D. Troisi, and Ariana F. Young. "Effects of Everyday Romantic Goal Pursuit on Women's Attitudes Toward Math and Science." *Personality and Social Psychology Bulletin*. May 26, 2011. Available at buffalo.edu/news/pdf/August11/ParkRomanticAttitudes.pdf.

[39] National Center for Education Statistics. *Trends in International Math and Science Study (TIMMS)*. "Highlights from TIMMS 2007: Mathematics and Science Achievement of U.S. Fourth- and Eighth-Grade Students in an International Context." September 2009. Available at nces.ed.gov/pubs2009/2009001.pdf.

[40] National Center for Education Statistics. *Program for International Student Assessment (PISA)*. "Highlights from PISA 2009: Performance of U.S. 15-Year-Old Students in Reading, Mathematics, and Science Literacy in an International Context." December 2010. Available at nces.ed.gov/pubs2011/2011004.pdf.

[41] National Science Foundation, National Science Board. "Science and Engineering Indicators 2010." NSB 10-01. 2010. Available at nsf.gov/statistics/seind10/pdf/seind10.pdf.

[42] World Economic Forum. "Global Competitiveness Report 2009-10." 2009. Available at members.weforum.org/pdf/GCR09/Report/Countries/United%20States.pdf.

[43] Greenstone, Michael, Adam Looney, and Paige Shevlin. "Improving Student Outcomes: Restoring America's Education Potential." The Hamilton Project, Brookings. September 2011. Available at brookings.edu/~/media/Files/rc/papers/2011/09_education_greenstone_looney_shevlin/092011_education_greens tone_looney_shevlin.pdf.

[44] National Center for Education Statistics. "The Nation's Report Card: Mathematics 2011: National Assessment of Education Progress at Grades 4 and 8." November 2011. Available at nces.ed.gov/nationsreportcard/pdf/main2011/2012458.pdf.

[45] *Ibid*.

[46] National Center for Education Statistics. "The Nation's Report Card: Science 2009: National Assessment of Education Progress at Grades 4, 8, and 12." January 2011. Available at nces.ed.gov/nationsreportcard/pdf/main2009/2011451.pdf.

[47] NAEP 2011 Math Results, *supra* note 44. In the 2011 math examination for fourth graders, there was a 25 point score gap between white and black students, and a 20 point score gap between white and Hispanic students; *See also* NAEP 2009 Science Results, *supra* note 46. On the 2009 science examination, the white-black student gaps in fourth, eighth, and twelfth grades were 36 points, 36 points, and 34 points, respectively, whereas the white-Hispanic student gaps were 32 points, 30 points, and 25 points.

[48] *Ibid*; *see also* NAEP 2011 Math Results, *supra* note 44.

[49] Leachman, Michael, Nicholas Johnson, and Erica Williams. "State Budget Cuts in the New Fiscal Year Are Unnecessarily Harmful." Center on Budget and Policy Priorities. July 28, 2011. Available at cbpp.org/cms/index.cfm?fa=view&id=3550.

[50] Bryant, Jeff. "Starving America's Public Schools: How Budget Cuts and Policy Mandates are Hurting our Nation's Students." Campaign for America's Future and the National Education Association. Available at ourfuture.org/files/documents/starving-schools-report.pdf.

INDEX

#

A

B

C

F

G

H

I

J

K

L

M

N

O

P

Q

R

S

Y